LONDON MATHEMATICAL SOCIETY STUDENT TEXTS

Managing editor: Dr C.M. Series, Mathematics Institute University of Warwick, Coventry CV4 7AL, United Kingdom

The books in the series listed below are available from booksellers, or, in case of difficulty, from Cambridge University Press.

1 Introduction to combinators and λ-calculus, J.R. HINDLEY & J.P. SELDIN
2 Building models by games, WILFRID HODGES
3 Local fields, J.W.S. CASSELS
4 An introduction to twistor theory, S.A. HUGGETT & K.P. TOD
5 Introduction to general relativity, L.P. HUGHSTON & K.P. TOD
6 Lectures on stochastic analysis: diffusion theory, DANIEL W. STROOCK
7 The theory of evolution and dynamical systems, J. HOFBAUER & K. SIGMUND
8 Summing and nuclear norms in Banach space theory, G.J.O. JAMESON
9 Automorphisms of surfaces after Nielsen and Thurston, A. CASSON & S. BLEILER
10 Nonstandard analysis and its applications, N. CUTLAND (ed)
11 Spacetime and singularities, G. NABER
12 Undergraduate algebraic geometry, MILES REID
13 An introduction to Hankel operators, J.R. PARTINGTON
14 Combinatorial group theory: a topological approach, DANIEL E. COHEN
15 Presentations of groups, D.L. JOHNSON
16 An introduction to noncommutative Noetherian rings, K.R.GOODEARL & R.B. WARFIELD, JR.
17 Aspects of quantum field theory in curved spacetime, S.A. FULLING
18 Braids and coverings: selected topics, Vagn Lundsgaard Hansen
19 Steps in commutative algebra, R.Y. SHARP
20 Communication Theory, C.M. GOLDIE & R.G.E. PINCH
21 Representations of finite groups of Lie type, FRANCOIS DIGNE & JEAN MICHEL
22 Designs, codes and graphs and their links, P. CAMERON & J. VAN LINT
23 Complex algebraic surfaces, F. KIRWAN
24 Lectures on elliptic curves, J.W.S. CASSELS
25 Hyperbolic geometry, B. IVERSEN
26 Elementary theory of L-functions and Eisenstein series, H. HIDA
27 Hilbert space: compact operators and the trace theorem, J.R. RETHERFORD
28 Potential theory in the complex plane, T. RANSFORD
29 Undergraduate commutative algebra, M. Reid
32 Lectures on Lie groups and Lie algebras, R. CARTER, G. SEGAL & I. MACDONALD
33 A primer of algebaic D-modules, S.C. COUTINHO
34 Complex algebraic surfaces (2nd ed), A. BEAUVILLE

London Mathematical Society Student Texts. 34

Complex Algebraic Surfaces

Second Edition

Arnaud Beauville
Université Paris-Sud

Published by the Press Syndicate of the University of Cambridge
The Pitt Building, Trumpington Street, Cambridge CB2 1RP
40 West 20th Street, New York, NY 10011-4211, USA
10 Stamford Road, Oakleigh, Victoria 3166, Australia

Originally published in French as *Surfaces Algébriques Complexes*, Astérique **54** and

Translated by R. Barlow,
With assistance from N.I. Shepherd-Barron and M. Reid

First published in English by Cambridge University Press 1983 as *Complex Algebraic Surfaces*

Second edition first published 1996

Library of Congress cataloguing in publication data available
A catalogue record for this book is available from the British Library

ISBN 0 521 49510 5 hardback
ISBN 0 521 49842 2 paperback

Transferred to digital printing 2003

CONTENTS

INTRODUCTION

This book is a modified version of a course given at Orsay in 1976–77. The aim of the course was to give a comparatively elementary proof of the Enriques classification of complex algebraic surfaces, accessible to a student familiar with the basic language of algebraic geometry (divisors, differential forms, ...) as well as sheaf cohomology. I have, however, preferred to assume along the way various hard theorems from algebraic geometry, rather than resort to complicated and artificial proofs.

Here is an outline of the course. The first two chapters introduce the basic tools for the study of surfaces: in Chapter I we define the intersection form on the Picard group, and establish its properties; assuming the duality theorem we deduce the fundamental results (the Riemann–Roch theorem, the genus formula). Chapter II is devoted to the structure of birational maps; we show in particular that every surface is obtained from a minimal surface by a finite number of blow-ups. The chapter ends with Castelnuovo's contractibility criterion, which characterizes exceptional curves by their numerical properties.

The classification begins in Chapter III with ruled surfaces, that is, surfaces birational to $\mathbb{P}^1 \times C$. We show that (except in the rational case) their minimal models are $\mathbb{P}^1$-bundles over a base curve C, and we study their geometry. Chapter IV gives some examples of rational surfaces; we take a stroll through the huge menagerie collected by the geometers of the 19th century (the Veronese surface, del Pezzo surfaces, ...).

The next two chapters are perhaps the keystone of the classification; they give the characterization of ruled surfaces by their numerical properties – more precisely, by the vanishing of the 'plurigenera' P_n. Surfaces with $q = 0$ are treated in Chapter V, where we prove Castelnuovo's theorem: a surface with $q = P_2 = 0$ is rational. We deduce two important consequences: the structure of minimal rational surfaces and the uniqueness of the minimal model of a non-ruled surface. In Chapter VI we begin the study of surfaces with $q > 0$. We show without too much trouble that a surface having $p_g = 0$ and $q \geqslant 2$ is ruled; which leaves certain non-ruled surfaces with $p_g = 0$ and $q = 1$. According to an idea of Enriques one can classify these surfaces very precisely, and show that they have $P_{12} > 0$. Thus a surface is ruled if and only if $P_{12} = 0$ (Enriques' theorem).

Chapter VII, which is very short, introduces the Kodaira dimension κ, which is a convenient invariant for the classification of surfaces. Ruled surfaces are characterised by $\kappa = -\infty$; the three ensuing chapters deal with surfaces with $\kappa = 0, 1$, and 2. Surfaces with $\kappa = 0$ fall into four classes: $K3$ surfaces, Enriques surfaces, Abelian surfaces, and bielliptic surfaces. The bielliptic surfaces were already listed in Chapter VI, in the context of surfaces with $p_g = 0$ and $q = 1$; here we study $K3$ surfaces and Enriques surfaces, and give numerous examples.

In Chapter IX we show that surfaces with $\kappa = 1$ have a (not necessarily rational) pencil of elliptic curves; conversely we study those surfaces with such a pencil.

Finally Chaper X concerns surfaces with $\kappa = 2$, said to be of general type; although these surfaces are the most general, there is not very much that we can say about them. We have limited ourselves to giving some examples and proving Castelnuovo's inequality $\chi(\mathcal{O}_S) > 0$.

In Appendix A we sketch (without proof) the classification of surfaces in characteristic p, and in Appendix B that of complex compact surfaces. Appendix C indicates some of the new results (or new approaches to old results) which have been obtained since the first appearance of this book.

It is hard to claim any originality in a subject whose main theorems were proved at the turn of the century. I have been largely inspired by the existing literature, in particular by Shafarevich's seminar [Sh 2]; in a historical note at the end of each chapter I have tried to describe the origins of the principal results. The exercises indicate various possible extensions to the course.

NOTATION

By 'surface' we shall mean smooth projective surface over the field $\mathbb{C}$ of complex numbers. Let S be a surface, and D, D' two divisors on S. We write:

$D \equiv D'$ if D and D' are linearly equivalent

$\mathcal{O}_S(D)$: the invertible sheaf corresponding to D

$H^i(S, \mathcal{O}_S(D))$, or simply $H^i(D)$: the ith cohomology group of the sheaf $\mathcal{O}_S(D)$

$h^i(D) = \dim_{\mathbb{C}} H^i(D)$

$\chi(\mathcal{O}_S(D)) = h^0(D) - h^1(D) + h^2(D)$, the Euler–Poincaré characteristic of the sheaf $\mathcal{O}_S(D)$

$|D|$ = the set of effective divisors linearly equivalent to D
= the projective space corresponding to $H^0(D)$

K_S or K = 'the' canonical divisor = a divisor such that $\mathcal{O}_S(K) \cong \Omega_S^2$

$\operatorname{Pic} S$ = the group of divisors on S modulo linear equivalence
$\cong$ group of isomorphism classes of invertible sheaves

$NS(S)$ = the Néron–Severi group of S (I.10)

$\operatorname{Alb}(S)$ = the Albanese variety of S (see Chapter V)

$q(S)$ or $q = h^1(\mathcal{O}_S) = h^0(\Omega_S^1)$

$p_g(S)$ or $p_g = h^2(\mathcal{O}_S) = h^0(\mathcal{O}_S(K))$

$P_n(S)$ or $P_n = h^0(\mathcal{O}_S(nK))$ (for $n \geqslant 1$)

$b_i(S)$ or $b_i = \dim_{\mathbb{R}} H^i(S, \mathbb{R})$

$\chi_{\text{top}}(S) = \sum(-1)^i b_i(S)$

I

THE PICARD GROUP AND THE RIEMANN–ROCH THEOREM

Unless otherwise stated, we consider surfaces with their Zariski topology (the closed subsets are the algebraic subvarieties); 'sheaf' will mean 'coherent algebraic sheaf'. This is a matter of convention: Serre's general theorems ([GAGA]) give a bijection between algebraic and analytic coherent sheaves which preserves exactness, cohomology, etc. All our arguments with coherent algebraic sheaves will remain valid in the analytic context.

Fact I.1 The Picard group

Let S be a smooth variety. Recall that the Picard group of S, $\operatorname{Pic} S$, is the group of isomorphism classes of invertible sheaves (or of line bundles) on S. To every effective divisor D on S there corresponds an invertible sheaf $\mathcal{O}_S(D)$ and a section $s \in H^0(\mathcal{O}_S(D))$, $s \neq 0$, which is unique up to scalar multiplication, such that $\operatorname{div}(s) = D$. We define $\mathcal{O}_S(D)$ for an arbitrary D by linearity. The map $D \mapsto \mathcal{O}_S(D)$ identifies $\operatorname{Pic} S$ with the group of linear equivalence classes of divisors on S.

Let X be another smooth variety and $f : S \to X$ a morphism. We can define the inverse image with respect to f of an invertible sheaf, to get a homomorphism $f^* : \operatorname{Pic} X \to \operatorname{Pic} S$. If f is surjective, then we can also define the inverse image of a divisor, in such a way that $f^*\mathcal{O}_X(D) = \mathcal{O}_S(f^*(D))$: just note that the inverse image of a non-zero section of $\mathcal{O}_X(D)$ is non-zero.

If f is a morphism of surfaces which is generically finite of degree d, then we define the direct image f_*C of an irreducible curve C by setting

$f_*C = 0$	if $f(C)$ is a point,
$f_*C = r\Gamma$	if $f(C)$ is a curve Γ, the morphism $C \to \Gamma$ induced by f being finite of degree r.

We define f_*D for all divisors D on S by linearity. One can check immediately that $D \equiv D'$ implies $f_*D \equiv f_*D'$. It follows from the definition that

$$f_*f^*D = dD \quad \text{for all divisors } D \text{ on } S\ .$$

The particular importance of the Picard group in the case of surfaces stems from the existence of an intersection form, which we now define.

Definition I.2 *Let C, C' be two distinct irreducible curves on a surface S, $x \in C \cap C'$, $\mathcal{O}_x$ the local ring of S at x. If f (resp. g) is an equation of C (resp. C') in $\mathcal{O}_x$, the intersection multiplicity of C and C' at x is defined to be*

$$m_x(C \cap C') = \dim_{\mathbb{C}} \mathcal{O}_x/(f,g)\ .$$

By the Nullstellensatz the ring $\mathcal{O}_x/(f,g)$ is a finite-dimensional vector space over $\mathbb{C}$. The reader should confirm that this definition corresponds to the intuitive notion of intersection number. For example, we see that $m_x(C \cap C') = 1$ if and only if f and g generate the maximal ideal ${}_x$, i.e. form a system of local coordinates in a neighbourhood of x : C and C' are then said to be transverse at x.

Definition I.3 *If C, C' are two distinct irreducible curves on S, the intersection number $(C.C')$ is defined by:*

$$(C.C') = \sum_{x \in C \cap C'} m_x(C \cap C')\ .$$

Recall that the ideal sheaf defining C (resp. C') is just the invertible sheaf $\mathcal{O}_S(-C)$ (resp. $\mathcal{O}_S(-C')$); define

$$\mathcal{O}_{C \cap C'} = \mathcal{O}_S/(\mathcal{O}_S(-C) + \mathcal{O}_S(-C'))\ .$$

The sheaf $\mathcal{O}_{C\cap C'}$ is a skyscraper sheaf, concentrated at the finite set $C \cap C'$; at each of these points we have $(\mathcal{O}_{C\cap C'})_x = \mathcal{O}_x/(f,g)$ (with the notation of I.1). It is now clear that

$$(C.C') = \dim H^0(S, \mathcal{O}_{C\cap C'})\ .$$

For any sheaf L on S, let $\chi(L) = \sum_i (-1)^i h^i(S,L)$ be the Euler–Poincaré characteristic of L.

Theorem I.4 *For L, L' in $\operatorname{Pic} S$, define*

$$(L.L') = \chi(\mathcal{O}_S) - \chi(L^{-1}) - \chi(L'^{-1}) + \chi(L^{-1} \otimes L'^{-1})\ .$$

Then $(\ .\ )$ *is a symmetric bilinear form on* $\operatorname{Pic} S$, *such that if* C *and* C' *are two distinct irreducible curves on* S *then*

$$(\mathcal{O}_S(C).\mathcal{O}_S(C')) = (C.C') \ .$$

Proof We start by proving the last equality.

Lemma I.5 *Let* $s \in H^0(S, \mathcal{O}_S(C))$ *(resp.* $s' \in H^0(S, \mathcal{O}_S(C'))$*) be a non-zero section vanishing on* C *(resp.* C'*). The sequence*

$$0 \to \mathcal{O}_S(-C-C') \xrightarrow{(s',-s)} \mathcal{O}_S(-C) \oplus \mathcal{O}_S(-C') \xrightarrow{(s,s')} \mathcal{O}_S \longrightarrow \mathcal{O}_{C\cap C'} \to 0$$

is exact.

Proof Let $f, g \in \mathcal{O}_x$ be local equations for C, C' at a point $x \in S$; we must show that the sequence

$$0 \to \mathcal{O}_x \xrightarrow{(g,-f)} \mathcal{O}_x^2 \xrightarrow{(f,g)} \mathcal{O}_x \to \mathcal{O}_x/(f,g) \to 0$$

is exact, i.e. that if $a, b \in \mathcal{O}_x$ are such that $af = bg$, then there exists $k \in \mathcal{O}_x$ such that $a = kg$, $b = kf$.

This follows immediately, say from the fact that $\mathcal{O}_x$ is a factorial ring and f, g are coprime (otherwise C and C' would have a common component). The highbrow reader can use the (much weaker) fact that $\mathcal{O}_x$ is Cohen–Macaulay.

Lemma I.5 and the additivity of the Euler–Poincaré characteristic give $(\mathcal{O}_S(C).\mathcal{O}_S(C')) = (C.C')$. To prove the theorem it remains to show that $(\ .\ )$ is a bilinear form on $\operatorname{Pic} S$ (the symmetry is obvious).

Lemma I.6 *Let* C *be a non-singular irreducible curve on* S. *For all* $L \in \operatorname{Pic} S$, *we have*

$$(\mathcal{O}_S(C).L) = \deg(L_{|C}) \ .$$

Proof The exact sequences

$$\begin{array}{ccccccccc} 0 & \to & \mathcal{O}_S(-C) & \to & \mathcal{O}_S & \to & \mathcal{O}_C & \to & 0 \\ 0 & \to & L^{-1}(-C) & \to & L^{-1} & \to & L^{-1} \otimes \mathcal{O}_C & \to & 0 \end{array}$$

give the following relations between Euler–Poincaré characteristics:

$$\begin{aligned} \chi(\mathcal{O}_S) - \chi(\mathcal{O}_S(-C)) &= \chi(\mathcal{O}_C) \\ \chi(L^{-1}) - \chi(L^{-1}(-C)) &= \chi(L^{-1}_{|C}) \end{aligned}$$

whence

$$\begin{aligned}(\mathcal{O}_S(C).L) &= \chi(\mathcal{O}_C) - \chi(L_{|C}^{-1}) \\ &= -\deg L_{|C}^{-1} \quad \text{by the Riemann–Roch theorem on } C \\ &= \deg L_{|C}, \qquad \text{which proves the lemma.}\end{aligned}$$

For L_1, L_2, $L_3 \in \operatorname{Pic} S$, consider the expression

$$s(L_1, L_2, L_3) = (L_1.L_2 \otimes L_3) - (L_1.L_2) - (L_1.L_3) .$$

It is clear by definition of the product that this is symmetric in L_1, L_2 and L_3; moreover Lemma I.6 shows $s(L_1, L_2, L_3)$ is zero when $L_1 = \mathcal{O}_S(C)$, with C a non-singular curve. Similarly, $s(L_1, L_2, L_3)$ is zero if L_2 or L_3 is of this form.

For the general case, we recall a theorem of Serre (cf. [FAC]):

Fact I.7 *Let D be a divisor on S, and H a hyperplane section of S (for a given embedding). There exists $n \geqslant 0$ such that $D + nH$ is a hyperplane section (for another embedding). In particular we can write $D \equiv A - B$, where A and B are smooth curves on S, with $A \equiv D + nH$ and $B \equiv nH$.*

Now let L, L' be any two invertible sheaves. We can write $L' = \mathcal{O}_S(A - B)$, where A and B are two smooth curves on S. Noting that $s(L, L', \mathcal{O}_S(B)) = 0$, we get

$$(L, L') = (L.\mathcal{O}_S(A)) - (L.\mathcal{O}_S(B)) .$$

Via Lemma I.6, we deduce that $(L.L')$ is linear in L. This completes the proof of Theorem I.4.

If D, D' are two divisors on S, we write $D.D'$ for $(\mathcal{O}_S(D).\mathcal{O}_S(D'))$. The point of Theorem I.4 is that we can calculate this product by replacing D (or D') by a linearly equivalent divisor. Here are two applications of this principle.

Proposition I.8

(i) *Let C be a smooth curve, $f : S \to C$ a surjective morphism, F a fibre of f. Then $F^2 = 0$.*

(ii) *Let S' be a surface, $g : S \to S'$ a generically finite morphism of degree d, D and D' divisors on S. Then $g^*D.g^*D' = d(D.D')$.*

Proof (i) $F = f^*[x]$, for some $x \in C$. There exists a divisor A on C, linearly equivalent to x, such that $x \notin A$, so that $F \equiv f^*A$. Since

f^*A is a linear combination of fibres of f all distinct from F, we have $F^2 = F.f^*A = 0$.

(ii) As in I.7, it is enough to prove the formula when D and D' are hyperplane sections of S (in two different embeddings). There exists an open set U in S' over which g is étale. We can move D and D' so that they meet transversely and their intersection lies in U. It is then clear that g^*D and g^*D' meet transversely and that $g^*D \cap g^*D' = g^{-1}(D \cap D')$; hence the result.

Examples I.9

(a) $S = \mathbb{P}^2$.
$\operatorname{Pic} \mathbb{P}^2 = \mathbb{Z}$: every curve of degree d is linearly equivalent to d times a line. Let C, C' be curves of degrees d, d', and L, L' distinct lines; since $C \equiv dL$, $C' \equiv d'L'$, Theorem I.4 gives Bezout's theorem:

$$C.C' = dL.dL' = dd'(L.L') = dd' .$$

(b) $S = \mathbb{P}^1 \times \mathbb{P}^1 =$ smooth quadric in $\mathbb{P}^3$ – because of the Segre embedding $\mathbb{P}^1 \times \mathbb{P}^1 \hookrightarrow \mathbb{P}^3$, defined by

$$((X,T);(X',T')) \mapsto (XX',XT',TX',TT').$$

Let $h_1 = \{0\} \times \mathbb{P}^1$, $h_2 = \mathbb{P}^1 \times \{0\}$, $U = (\mathbb{P}^1 \times \mathbb{P}^1) - h_1 - h_2$. The open set U is isomorphic to the affine space $\mathbb{A}^2$, so every divisor on U is the divisor of a rational function. Let D be a divisor on $\mathbb{P}^1 \times \mathbb{P}^1$; then $D_{|U} = \operatorname{div}(\phi)$ on U, so there exist integers n and m such that

$$D = \operatorname{div}(\phi) + mh_1 + nh_2$$

and $D \equiv mh_1 + nh_2$.

Thus $\operatorname{Pic} S$ is generated by the classes of h_1 and h_2. It is clear that $h_1.h_2 = 1$. To find h_1^2, by I.4, we can replace h_1 by the curve $C_\infty = \{\infty\} \times \mathbb{P}^1$ which is linearly equivalent to h_1; since $h_1 \cap C_\infty = \emptyset$, we have $h_1^2 = h_1.C_\infty = 0$. Similarly $h_2^2 = 0$. It follows that $\mathcal{O}_S(h_1)$ and $\mathcal{O}_S(h_2)$ generate $\operatorname{Pic} S$, and the intersection product is given by $h_1^2 = h_2^2 = 0$, $h_1.h_2 = 1$.

Let Γ be a curve in $\mathbb{P}^1 \times \mathbb{P}^1$; it is defined by a bihomogeneous equation, i.e. homogeneous of weight m in the coordinates (X,T) and of weight n in (X',T'); Γ is said to have bidegree (m,n), and

$\Gamma \equiv mh_1 + nh_2$. If Γ' is a curve of bidegree (m', n'), Theorem I.4 gives:

$$\Gamma.\Gamma' = (mh_1 + nh_2).(m'h_1 + n'h_2) = mn' + nm' .$$

I.10 Topological interpretation Here we use analytic sheaves on S, and the ordinary topology. We write ${}^h\mathcal{O}_S$ for the sheaf of holomorphic functions on S, considered as an analytic manifold.

Consider the exponential map $e : {}^h\mathcal{O}_S \to {}^h\mathcal{O}_S^*$, given by $e(f) = \exp(2\pi i f)$. It is locally surjective (local existence of logarithms), and its kernel clearly consists of locally constant integer-valued functions. In other words there is an exact sequence

$$0 \to \mathbb{Z} \to {}^h\mathcal{O}_S \xrightarrow{e} {}^h\mathcal{O}_S^* \to 1 .$$

Let us study the derived long exact cohomology sequence. Since $H^0(S, {}^h\mathcal{O}_S) = \mathbb{C}$ and $H^0(S, {}^h\mathcal{O}_S^*) = \mathbb{C}^*$, we can start with the H^1:

$$0 \to H^1(S,\mathbb{Z}) \to H^1(S, {}^h\mathcal{O}_S) \to H^1(S, {}^h\mathcal{O}_S^*) \to H^2(S,\mathbb{Z}) \to H^2(S, {}^h\mathcal{O}_S) .$$

We know that $H^1(S, \mathcal{O}_S^*)$ is canonically identified with $\operatorname{Pic} S$ (strictly speaking the analytic Pic, but from [GAGA] this is the same thing; likewise $H^i(S, {}^h\mathcal{O}_S) \cong H^i(S, \mathcal{O}_S)$). So the group $\operatorname{Pic} S$ appears as an extension:

$$0 \to T \to \operatorname{Pic} S \to NS(S) \to 0$$

of two groups which are different in nature: $T = H^1(S, \mathcal{O}_S)/H^1(S, \mathbb{Z})$ is a divisible group (Hodge theory shows that $H^1(S, \mathbb{Z})$ is a lattice in $H^1(S, \mathcal{O}_S)$, so T has a natural structure of complex torus); whereas $NS(S) \subset H^2(S, \mathbb{Z})$ is a finitely generated group, called the Néron–Severi group of S.

The map $c : \operatorname{Pic} S \to H^2(S, \mathbb{Z})$ can be described topologically as follows. If $C \subset S$ is an irreducible curve, the restriction $H^2(S, \mathbb{Z}) \to H^2(C, \mathbb{Z}) \cong \mathbb{Z}$ gives a linear form on $H^2(S, \mathbb{Z})$, hence by Poincaré duality an element $c(C) \in H^2(S, \mathbb{Z})$; we define $c(D)$ for any divisor D by linearity. Then $c(D).c(D') = D.D'$ for divisors D, D' on S; in other words, the intersection form comes from the bilinear form on $NS(S)$ induced by the cup product.

If f is a morphism from S to a smooth variety X, then $f^*c(S) = c(f^*D)$ for any divisor D on X. If X is a surface, the Gysin homomorphism $f_* : H^2(S, \mathbb{Z}) \to H^2(X, \mathbb{Z})$ is defined, and $f_*c(D) = c(f_*D)$ for any D on S.

Let us come back to our algebraic set-up. We recall without proof Serre duality:

Theorem I.11 *Let S be a surface, and L a line bundle on S. Let ω_S be the line bundle of differential 2-forms on S. Then $H^2(S,\omega_S)$ is a 1-dimensional vector space; for $0 \leqslant i \leqslant 2$, the cup-product pairing*

$$H^i(S,L) \otimes H^{2-i}(S,\omega_S \otimes L^{-1}) \to H^2(S,\omega_S) \xrightarrow{\sim} \mathbb{C}$$

defines a duality. In particular, $\chi(L) = \chi(\omega_S \otimes L^{-1})$.

We can now prove the Riemann–Roch theorem.

Theorem I.12 (Riemann–Roch) *For all $L \in \operatorname{Pic} S$,*

$$\chi(L) = \chi(\mathcal{O}_S) + \frac{1}{2}(L^2 - L.\omega_S) .$$

Proof Let us compute $(L^{-1}.L \otimes \omega_S^{-1})$. By definition of the intersection product

$$(L^{-1}.L \otimes \omega_S^{-1}) = \chi(\mathcal{O}_S) - \chi(L) - \chi(\omega_S \otimes L^{-1}) + \chi(\omega_S) .$$

By Serre duality, $\chi(\omega_S) = \chi(\mathcal{O}_S)$ and $\chi(\omega_S \otimes L^{-1}) = \chi(L)$, and hence

$$(L^{-1}.L \otimes \omega_S^{-1}) = 2(\chi(\mathcal{O}_S) - \chi(L)) .$$

Using the bilinearity of the intersection form gives at once the stated formula.

Remarks I.13

(i) We will usually be writing these two theorems in terms of divisors; set $h^i(D) = h^i(S,\mathcal{O}_S(D))$; furthermore, it is traditional to let K denote any divisor such that $\mathcal{O}_S(K) = \omega_S$, and to call K a 'canonical divisor'. Serre duality then gives $h^i(D) = h^{2-i}(K-D)$ for $0 \leqslant i \leqslant 2$; and the Riemann–Roch theorem takes the form

$$h^0(D) + h^0(K-D) - h^1(D) = \chi(\mathcal{O}_S) + \frac{1}{2}(D^2 - D.K) .$$

Usually we will not have any information about $h^1(D)$, and we will use Riemann–Roch as an inequality

$$H^0(D) + H^0(K-D) \geqslant \chi(\mathcal{O}_S) + \frac{1}{2}(D^2 - D.K) .$$

(ii) We have given the Riemann–Roch theorem in its classical form. Nowadays one usually takes the Riemann–Roch theorem to mean Hirzebruch's version, which is the conjunction of I.12 and of the important formula of M. Noether, which we will assume:

I.14 Noether's formula

$$\chi(\mathcal{O}_S) = \frac{1}{12}(K_S^2 + \chi_{\text{top}}(S)) ,$$

where $\chi_{\text{top}}(S)$ *is the topological Euler–Poincaré characteristic of* S: $\chi_{\text{top}}(S) = \sum(-1)^i b_i$, with $b_i = \dim_{\mathbb{R}} H^i(S, \mathbb{R})$.

Here is an important consequence of the Riemann–Roch formula.

I.15 The genus formula *Let* C *be an irreducible curve on a surface* S. *The genus of* C, *defined by* $g(C) = h^1(C, \mathcal{O}_C)$, *is given by* $g(C) = 1 + \frac{1}{2}(C^2 + C.K)$.

Proof The exact sequence

$$0 \to \mathcal{O}_S(-C) \to \mathcal{O}_S \to \mathcal{O}_C \to 0$$

gives $\chi(\mathcal{O}_C) = 1 - g(C) = \chi(\mathcal{O}_S) - \chi(\mathcal{O}_S(-C))$; the formula then follows from Riemann–Roch.

Remarks I.16

(i) Note that the genus of C is not the same as that of its normalization. More precisely, let $f : N \to C$ denote the normalization of C; we define a sheaf δ on C by the exact sequence

$$0 \to \mathcal{O}_C \to f_*\mathcal{O}_N \to \delta \to 0.$$

The cokernel δ is concentrated at the singular points of C, so that $H^1(C, \delta) = 0$, and $H^0(C, \delta) = \bigoplus_{x \in C} \delta_x$; $\delta = 0$ if and only if f is an isomorphism, that is C is smooth.

The associated cohomology exact sequence gives

$$g(C) = g(N) + \sum_{x \in C} \dim(\delta_x) .$$

Hence $g(C) > g(N)$ if C is singular; in particular, the condition $g(C) = 0$ implies that C is rational and smooth, that is $C \cong \mathbb{P}^1$.

(ii) The genus formula can also be written $2g-2 = \deg(K+C)_{|C}$ (see I.6). If C is smooth, we will now show that in fact $(\mathcal{O}_S(K+C))_{|C} = \omega_C$ (*adjunction formula*). For this, recall:

Fact I.17 *Let X and Y be two smooth varieties, and $j : X \hookrightarrow Y$ an embedding; write I for the ideal of X in Y. The sheaf $j^*I = I/I^2$ is then locally free of rank codim(X,Y) on X, and we have an exact sequence*

$$0 \to I/I^2 \xrightarrow{d} j^*\Omega^1_Y \xrightarrow{j^*} \Omega^1_X \to 0 .$$

Going back to the case $C \subset S$, we have $I = \mathcal{O}_S(-C)$, and this sequence becomes

$$0 \to \mathcal{O}_S(-C)_{|C} \to \Omega^1_{S|C} \to \omega_C \to 0 .$$

Considering the exterior powers gives the claimed equality.
(If C is singular, we still have $(\mathcal{O}_S(K+C))_{|C} = \omega_C$, where ω_C is the dualizing sheaf of C. But we will not use duality theory for singular curves.)

Historical Note I.18
The material of this chapter is the foundation of the theory of surfaces; it was all known before 1900. Linear systems, well understood on curves, were introduced in complete generality on sufaces by Max Noether ([N1]), and later studied very fully by Enriques ([E1]). The genus formula was proved by Noether in 1886 ([N2]), who used it to deduce the Riemann–Roch theorem, although assuming implicitly that $h^1(D) = h^1(\mathcal{O}_S) = 0$. The right form was given by Enriques in 1896 ([E1]), based on a result of Castelnuovo.

Noether's formula was proved in [N1]: Noether projects the surface birationally onto a singular surface in $\mathbb{P}^3$, and explicitly calculates the 3 invariants appearing in the formula; χ_{top} appears in the guise of the 'Zeuthen-Segre invariant'.

These geometers only considered effective curves; but the need to introduce 'virtual curves', that is, divisors, was quickly felt, especially by Severi. The theory is then complete. However, Serre's introduction of coherent sheaves in 1955 ([FAC]) revolutionized the presentation, turning most of the results into formalities. In 1956, Hirzebruch generalized the Riemann–Roch theorem to varieties of arbitrary dimension ([H]). His version contains Noether's formula; let us give a brief indication of his proof. A formal computation of characteristic classes gives

$p_1 = K^2 - 2\chi_{\mathrm{top}}(S)$, where p_1 is the first Pontryagin class of S. Cobordism theory then shows that $p_1 = 3\tau$, where τ is the signature of the intersection form on $H^2(S,\mathbb{Z})$; indeed, the two sides of this equation are cobordism invariants, and they agree on $\mathbb{P}^2$. Finally, Hodge theory gives $\tau = 2 + 4h^2(\mathcal{O}_S) - b_2 = 4\chi(\mathcal{O}_S) - \chi_{\mathrm{top}}(S)$. We deduce

$$K^2 + \chi_{\mathrm{top}}(S) = 3\tau + 3\chi_{\mathrm{top}}(S) = 12\chi(\mathcal{O}_S)\ .$$

Our presentation of the intersection form follows essentially [M1].

II
BIRATIONAL MAPS

Before beginning a classification, we have to decide when we are going to consider two of the objects we are classifying to be equivalent. In algebraic geometry, we can classify varieties up to isomorphism or, more coarsely, up to birational equivalence. The problem does not arise for curves, since a rational map from one smooth complete curve to another is in fact a morphism. For surfaces, we shall see that the structure of birational maps is very simple; they are composites of 'elementary' birational maps, the blow-ups. We should point out that the problem is much more complex in higher dimensions.

II.1 Blow-ups

Let S be a surface and $p \in S$. Then there exist a surface $\hat{S}$ and a morphism $\epsilon : \hat{S} \to S$, which are unique up to isomorphism, such that

(i) *the restriction of ϵ to $\epsilon^{-1}(S - \{p\})$ is an isomorphism onto $S - \{p\}$;*

(ii) *$\epsilon^{-1}(p) = E$, say, is isomorphic to $\mathbb{P}^1$.*

We shall say that ϵ is the blow-up of S at p, and E is the exceptional curve of the blow-up.

Let us run rapidly over the construction of ϵ. Take a neighbourhood U of p on which there exist local coordinates x, y at p (i.e. the curves $x = 0$, $y = 0$ intersect transversely at p). We can assume that p is the only point of U in the intersection of these two curves. Define the subvariety $\hat{U}$ of $U \times \mathbb{P}^1$ by the equation $xY - yX = 0$, where X, Y are homogeneous coordinates on $\mathbb{P}^1$. It is clear that the projection $\epsilon : \hat{U} \to U$ is an isomorphism over the points of U where at most one of the coordinates x, y vanishes, while $\epsilon^{-1}(p) = \{p\} \times \mathbb{P}^1$. We get S by passing $\hat{U}$ and $S - \{p\}$ along $U - \{p\} \cong \hat{U} - \epsilon^{-1}(p)$.

It follows from this construction that the points of E can be naturally identified with the tangent directions on S at p. Let $\epsilon : \hat{S} \to S$ be the blow-up of a point p, and consider an irreducible curve C on S that passes through p with multiplicity m. The closure of $\epsilon^{-1}(C - \{p\})$ in $\hat{S}$ is an irreducible curve $\hat{C}$ on $\hat{S}$, which we call the strict transform of C.

Lemma II.2 $\epsilon^* C = \hat{C} + mE$.

Proof It is clear that $\epsilon^* C = \hat{C} + kE$ for some $k \in \mathbb{N}$. Choose local coordinates x, y in a neighbourhood U of p such that the curve $y = 0$ is not tangent to any branch of C at p. Then in the complete local ring $\hat{\mathcal{O}}_{S,p}$, the equation of C can be written as a formal power series

$$f = f_m(x,y) + f_{m+1}(x,y) + \cdots$$

where the f_k are homogeneous polynomials in x, y of degree k. The integer m is by definition the multiplicity of C at p, and we have $f_m(x,y) \neq 0$. Now construct $\hat{U} \subset U \times \mathbb{P}^1$ as above; in a neighbourhood of the point (p, ∞) of $\hat{U}$, we can take the functions x and $t = Y/X$ as local coordinates. Then

$$\epsilon^* f = f(x, tx) = x^m [f_m(1,t) + x.f_{m+1}(1,t) + \cdots] ,$$

and it follows at once that $k = m$.

Proposition II.3 *Let S be a surface, $\epsilon : \hat{S} \to S$ the blow-up of a point $p \in S$ and $E \subset \hat{S}$ the exceptional curve.*

(i) *There is an isomorphism* $\operatorname{Pic} S \oplus \mathbb{Z} \xrightarrow{\sim} \operatorname{Pic} \hat{S}$ *defined by* $(D, n) \mapsto \epsilon^* D + nE$.

(ii) *Let D, D' be divisors on S. Then $(\epsilon^* D).(\epsilon^* D') = D.D'$, $E.(\epsilon^* D) = 0$, $E^2 = -1$.*

(iii) *$NS(\hat{S}) \cong NS(S) \oplus \mathbb{Z}[E]$.*

(iv) *$K_{\hat{S}} = \epsilon^* K_S + E$.*

Proof (ii) To establish the first two formulae, we can replace D and D' by linearly equivalent divisors (Theorem I.4), and so suppose that p lies on no component of D nor D'. The two formulae are now obvious.

Lastly, choose a curve C passing through p with multiplicity 1. Its strict transform $\hat{C}$ meets E transversely at one point, which corresponds in E to the tangent direction defined at p by C. Thus $C.E = 1$; since $\hat{C} = \epsilon^* C - E$ (Lemma II.2) and $\epsilon^* C.E = 0$, we get $E^2 = -1$.

(i) Every irreducible curve on $\hat{S}$, other than E, is the strict transform of its image in S; it follows that the map defined is surjective. Suppose that there is a divisor D on S such that $\epsilon^*D + nE = 0$. Taking the intersection product with E, we see that $n = 0$, and upon applying ϵ_* we see that $D = 0$. This proves (i).

(iii) In view of the fact that ϵ_* and ϵ^* are defined on the Néron–Severi groups, the same proof gives (iii).

(iv) Choose a meromorphic 2-form ω on S such that ω is holomorphic in a neighbourhood of p and $\omega(p) \neq 0$. It is plain that away from E the zeros and poles of $\epsilon^*\omega$ are those of ω (via ϵ). Thus $\operatorname{div}(\epsilon^*\omega) = \epsilon^*\operatorname{div}(\omega) + kE$, and we must show that $k = 1$. This follows from the genus formula (I.15) and from what we have proved already. Alternatively, the reader can check that if $\omega = dx \wedge dy$, where x, y are local coordinates at $p \in S$, then $\epsilon^*\omega = x\, dx \wedge dt$ in local coordinates x, t at a point of $E \subset \hat{S}$.

We can now start to prove the fundamental results in the birational geometry of surfaces. We begin by recalling some facts.

II.4 Rational Maps

Let X, Y be varieties with X irreducible. A rational map $\phi : X \dashrightarrow Y$ is a morphism from an open subset U of X to Y which cannot be extended to any larger open subset. We say that ϕ is defined at x if $x \in U$.

Suppose that X is a (smooth) surface. Then $X - U = F$ is a finite set. (To see this, we can reduce at once to the case $Y = \mathbb{P}^n$ and then to $Y = \mathbb{P}^1$, i.e. to the case where ϕ is a rational function. Then the points where ϕ is not defined are just those lying in the intersection of the divisors of zeros and of poles of ϕ, and so form a finite set.) Here are two applications of this result:

(a) Let C be an irreducible curve on S; then ϕ is defined on $C - F$. Thus we can speak of the image of C, denoted by $\phi(C)$, by setting $\phi(C) = \overline{\phi(C - F)}$. Similarly we can write $\phi(S) = \overline{\phi(S - F)}$.

(b) Restriction induces an isomorphism between the divisor groups of S and of $S - F$, which induces an isomorphism $\operatorname{Pic} S \xrightarrow{\sim} \operatorname{Pic}(S - F)$. Thus we can speak of the inverse image under ϕ of a divisor D (or of a linear system P, or an invertible sheaf L) on Y, which we denote by ϕ^*D (resp. ϕ^*P, ϕ^*L).

II.5 Linear Systems

For a divisor D on a surface S, we denote by $|D|$ the set of all effective divisors on S linearly equivalent to D. Every non-vanishing section of $\mathcal{O}_S(D)$ defines an element of $|D|$, namely its divisor of zeros, and conversely every element of $|D|$ is the divisor of zeros of a non-vanishing section of $\mathcal{O}_S(D)$, defined up to scalar multiplication. Thus $|D|$ can be naturally identified with the projective space associated to the vector space $H^0(\mathcal{O}_S(D))$. A linear subspace P of $|D|$ is called a linear system on S; equivalently P can be defined by a vector subspace of $H^0(\mathcal{O}_S(D))$. We say P is complete if $P = |D|$. The dimension of P is by definition its dimension as a projective space. A 1-dimensional linear system is a pencil.

We say that a curve C is a fixed component of P if every divisor of P contains C. The fixed part of P is the biggest divisor F that is contained in every element of P. For any $D \in P$, the linear system $|D - F|$ has no fixed part.

A point p of S is said to be a base point or fixed point of P if every divisor of P contains p. If the linear system P has no fixed part, then it has only a finite number, say b, of fixed points; clearly $b \leqslant D^2$, for $D \in P$.

II.6 Rational Maps and Linear Systems

Let S be a surface. Then there is a bijection between the following sets:

(i) {rational maps $\phi : S \dashrightarrow \mathbb{P}^n$ such that $\phi(S)$ is contained in no hyperplane};

(ii) {linear systems on S without fixed part and of dimension n}.

This correspondence is constructed as follows: to the map ϕ we associate the linear system $\phi^*|H|$, where $|H|$ is the system of hyperplanes in $\mathbb{P}^n$ (see II.4). Conversely, let P be a linear system on S with no fixed part and denote by $\check{P}$ the projective space dual to P. Now define a rational map $\phi : S \dashrightarrow \check{P}$ by sending $x \in S$ to the hyperplane in P consisting of the divisors passing through x; ϕ is defined at x if and only if x is not a base point of P.

Theorem II.7 (elimination of indeterminacy) *Let $\phi : S \dashrightarrow X$ be a rational map from a surface to a projective variety. Then there exists a surface S', a morphism $\eta : S' \to S$ which is the composite of a finite number of blow-ups, and a morphism $f : S' \to X$ such that the diagram*

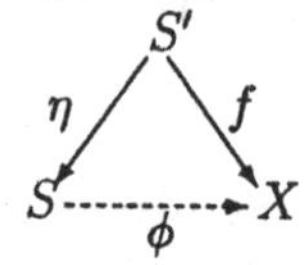

is commutative.

Proof Since X lies in some projective space, we may assume that $X = \mathbb{P}^m$. Moreover, we may suppose that $\phi(S)$ lies in no hyperplane of $\mathbb{P}^m$. Then ϕ corresponds to a linear system $P \subset |D|$ of dimension m on S, with no fixed component. If P has no base point, then ϕ is a morphism and there is nothing to prove.

Suppose then that P has a base point x, say, and consider the blow-up $\epsilon : S_1 \to S$ at x. Then the exceptional curve E occurs in the fixed part of the linear system $\epsilon^* P \subset |\epsilon^* D|$ with some multiplicity $k \geqslant 1$; in other words, the system $P_1 \subset |\epsilon^* D - kE|$, obtained by subtracting kE from each element of $\epsilon^* P$, has no fixed component. Hence it defines a rational map $\phi_1 : S_1 \dashrightarrow \mathbb{P}^m$, which is identical to $\phi \circ \epsilon$. If ϕ_1 is a morphism, then we are done; if not, we repeat the process. Thus we get by induction a sequence $\epsilon_n : S_n \to S_{n-1}$ of blow-ups and a linear system $P_n \subset |D_n|$ on S_n with no fixed part, where $D_n = \epsilon_n^* D_{n-1} - k_n E_n$. II.3(ii) gives $D_n^2 = D_{n-1}^2 - k_n^2 < D_{n-1}^2$; since P_n has no fixed part, $D_n^2 \geqslant 0$ for all n, and so this process must terminate. In other words, we arrive eventually at a system P_n with no base points, which defines a morphism $f : S_n \to \mathbb{P}^m$, as required.

Note that this proof gives an explicit construction of S' and f, and shows that D^2 is an upper bound on the number of blow-ups required.

Proposition II.8 (universal property of blowing-up) *Let $f : X \to S$ be a birational morphism of surfaces, and suppose that the rational map f^{-1} is undefined at a point p of S. Then f factorizes as*

$$f : X \xrightarrow{g} \hat{S} \xrightarrow{\epsilon} S$$

where g is a birational morphism and ϵ is the blow-up at p.

Lemma II.9 *Let S be an irreducible, but possibly singular, surface, S' a smooth surface and $f : S \to S'$ a birational morphism. Suppose that the rational map f^{-1} is undefined at a point $p \in S'$. Then $f^{-1}(p)$ is a curve on S.*

Proof We may suppose S to be affine (with $f^{-1}(p) \neq \emptyset$), so that there is an embedding $j : S \hookrightarrow \mathbb{A}^n$. The rational map $j \circ f^{-1} : S' \dashrightarrow \mathbb{A}^n$ is defined by rational functions $g_1, \ldots, g_n$, and one of them, say g_1, is undefined at x; i.e. $g_1 \notin \mathcal{O}_{S',p}$. Write $g_1 = u/v$, with $u, v \in \mathcal{O}_{S',p}$, u and v coprime, and $v(p) = 0$. Consider the curve D on S defined by $f^* v = 0$. On S, we have $f^* u = x_1 f^* v$, where x_1 is the first coordinate function on $S \subset \mathbb{A}^n$. It follows that $f^* u = f^* v = 0$ on D, so that $D = f^{-1}(Z)$, where Z is the subset of S' defined by $u = v = 0$. Since u, v are coprime, Z is finite; shrinking S' if necessary, we can assume that $Z = \{p\}$. Then $D = f^{-1}(p)$ as sets.

Lemma II.10 *Let $\phi : S \dashrightarrow S'$ be a rational map of surfaces such that ϕ^{-1} is undefined at a point $p \in S'$. Then there is a curve C on S such that $\phi(C) = \{p\}$.*

Proof The map ϕ corresponds to a morphism $f : U \to S'$ for some open subset $U \subset S$. Let $\Gamma \subset U \times S'$ denote the graph of f, i.e. the set $\{(u, f(u)) \mid u \in U\}$, and let S_1 denote the closure of Γ in $S \times S'$. S_1 is an irreducible surface, possibly with singularities. The projections q, q' of S_1 onto S, S' are birational morphisms and the diagram

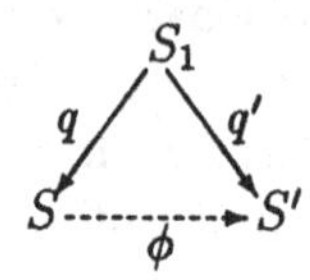

is commutative.

Since ϕ^{-1} is undefined at $p \in S'$, q'^{-1} is undefined at p. Then by Lemma II.9 there is an irreducible curve $C_1 \subset S_1$ with $q'(C_1) = \{p\}$. Since $S_1 \subset S \times S'$, $q(C_1)$ is a curve C in S with $\phi(C) = \{p\}$.

Proof of Proposition II Let g denote the birational map $\epsilon^{-1} \circ f$, and set $s = g^{-1}$. Suppose that g is undefined at a point $q \in X$. By Lemma II.10 there is a curve $C \subset \hat{S}$ such that $s(C) = \{q\}$; it follows that $\epsilon(C) = f(q)$, so that $C = E$ and $f(q) = p$. Let $\mathcal{O}_{X,q}$ be the local ring of X at q, and $_q$ its maximal ideal. There is a local coordinate y on S at p such that $f^*y \in {}^2_q$. Indeed, let (x, t) be a local coordinate system at p; if $g^*t \notin {}^2_q$, then it vanishes on $g^{-1}(p)$ with multiplicity 1, and so defines a local equation for $g^{-1}(p)$ in $\mathcal{O}_{X,q}$. Thus $g^*x = u\, g^*t$ for some $u \in \mathcal{O}_{X,q}$. Putting $y = x - u(q)t$, we see that $g^*y = (u - u(q))\, g^*t \in {}^2_q$.

So let e be any point of E where the map s is defined. We have $s^*g^*y = \epsilon^*y \in {}^2_e$, and this holds for all $e \in E$ outside some finite set. But it follows from the construction of the blow-up that ϵ^*y is a local coordinate at every point of E except one, which is absurd.

Theorem II.11 *Let $f : S \to S_0$ be a birational morphism of surfaces. Then there is a sequence of blow-ups $\epsilon_k : S_k \to S_{k-1}$ $(k = 1, \ldots, n)$ and an isomorphism $u : S \xrightarrow{\sim} S_n$ such that $f = \epsilon_1 \circ \cdots \circ \epsilon_n \circ u$.*

Proof If f is an isomorphism there is nothing to prove; otherwise there is a point p of S_0 such that f^{-1} is undefined at p. By Proposition II.8, f factorizes as $\epsilon_1 \circ f_1$, where ϵ_1 is the blow-up of S_0 at p and $f_1 : S \to S_1$ is a birational morphism. If the result fails, then by induction we can construct an infinite sequence of blow-ups $\epsilon_k : S_k \to S_{k-1}$ and birational morphisms $f_k : S \to S_k$ such that $\epsilon_k \circ f_k = f_{k-1}$. Denote by $n(f_k)$ the number of irreducible curves contracted to a point by f_k. Since $\epsilon_k \circ f_k = f_{k-1}$, it is clear that any curve contracted by f_k is also contracted by f_{k-1}; moreover, there is at least one irreducible curve C on S_k such that $f_k(C)$ is the exceptional curve of ϵ_k. Hence C is contracted by f_{k-1} but not by f_k, and so $n(f_k) < n(f_{k-1})$; thus $n(f_k) < 0$ for sufficiently large k, which is absurd.

Corollary II.12 *Let $\phi : S \dashrightarrow S'$ be a birational map of surfaces. Then there is a surface $\hat{S}$ and a commutative diagram*

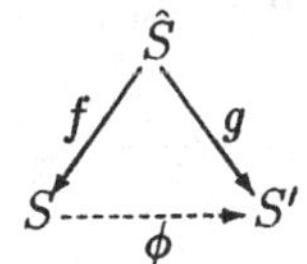

where the morphisms f, g are composites of blow-ups and isomorphisms.

Proof This follows at once from Theorems II.7 and II.11.

Remarks II.13

(1) Let $f : S \to S'$ be a birational morphism of surfaces which is a composite of n blow-ups (and an isomorphism). It follows from II.1(iii) that $NS(S) \cong NS(S') \oplus \mathbb{Z}^n$; since the Néron–Severi groups are finitely generated, we see that n is uniquely determined (i.e. independently of the factorization chosen). It also follows that every birational morphism from S to itself is an isomorphism.

(2) The blow-up $\epsilon : \hat{S} \to S$ at a point p has also a universal property 'in the other direction'; every morphism f from $\hat{S}$ to a variety X that contracts E to a point factors through S. The proof this time is very easy; we reduce first to the case X affine, then $X = \mathbb{A}^n$, then $X = \mathbb{A}^1$. Then f defines a function on $S - \{p\}$. But every function on $S - \{p\}$ extends over S, and the result follows.

II.14 Examples

(1) Let $S \subset \mathbb{P}^n$ be a surface and $p \in S$. The set of lines through p can be identified with a projective space $\mathbb{P}^{n-1}$. By associating to every point $q \in S - \{p\}$ the line $\langle q, p \rangle$ we get a rational map $S \dashrightarrow \mathbb{P}^{n-1}$, called the projection away from p. It is defined outside p, and one checks easily that it extends to a morphism $\hat{S} \to \mathbb{P}^{n-1}$, where $\hat{S}$ is the blow-up of S at p.

(2) Consider in particular a smooth quadric $Q \subset \mathbb{P}^2$. Projection from a point $p \in Q$ defines a morphism $f : \hat{Q} \to \mathbb{P}^2$. The inverse image of a point of $\mathbb{P}^2$, corresponding to a line ℓ passing through p, consists of the other point of intersection of ℓ with Q - except if ℓ lies in Q, in which case it is the whole of ℓ. Hence f is a birational morphism which contracts the two generators of Q passing through p (their strict transforms in $\hat{Q}$ are disjoint). Thus we have a commutative diagram

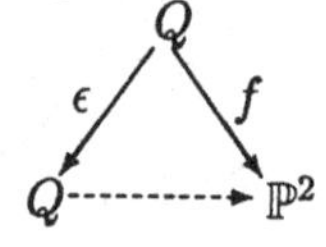

where f is (up to isomorphism) the blow-up of $\mathbb{P}^2$ at two distinct points, and ϵ is the blow-up of Q at p.

(3) Let p, q, r be three non-collinear points of $\mathbb{P}^2$ and P the linear system (of dimension 2) of conics passing through these points. P defines a rational map $\phi : \mathbb{P}^2 \dashrightarrow \check{P} \cong \mathbb{P}^2$, called a quadratic transformation. The system P has three base points p, q, r and one can check that ϕ extends to a morphism $f : S \to \check{P}$, where S is the blow-up of $\mathbb{P}^2$ at p, q, r. Let $x \in \check{P}$, corresponding to a pencil of conics $\{\lambda C_1 + \mu C_2\}$, where C_1, $C_2 \in P$; $f^{-1}(x)$ contains the base points of this pencil on S, namely the points of $\hat{C}_1 \cap \hat{C}_2$, where $\hat{C}_i$ is the strict transform of C_i. There are two possibilities:

(a) $C_1 \cap C_2$ consists of four points, p, q, r and a fourth point s, which is the only point of $f^{-1}(x)$.

(b) C_1 and C_2 have a common line ℓ; then $C_1 \cap C_2 = \ell \cup \{t\}$, where $t \in \mathbb{P}^2$. This can only happen if t is one of p, q, r and ℓ passes through the other two, and so there are three such pencils x, y, $z \in \check{P}$. Hence ϕ is birational and there is a commutative diagram

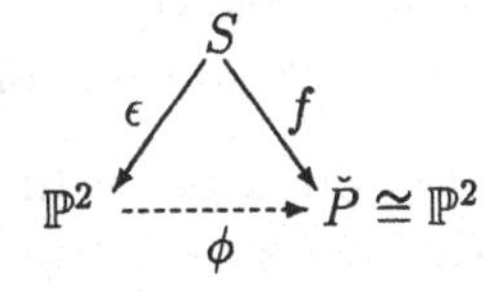

where ϵ is the blow-up of $\mathbb{P}^2$ at p, q, r, and f is, up to isomorphism, the blow-up of $\check{P}$ at x, y, z. The reader should check that the map $\mathbb{P}^2 \dashrightarrow \mathbb{P}^2$ given by

$$(X, Y, Z) \mapsto (X^{-1}, Y^{-1}, Z^{-1})$$

is of this type.

We can now compare the problem of the birational classification of surfaces with that of classification up to isomorphism. For a surface S let $B(S)$ denote the set of isomorphism classes of surfaces birationally equivalent to S. If S_1, $S_2 \in B(S)$, then S_1 is said to dominate S_2 if there is a birational morphism $S_1 \to S_2$. In view of Remark II.13.(1)), we can define an order on $B(S)$.

Definition II.15 *A surface S is minimal if its class in $B(S)$ is minimal, so that every birational morphism $S \to S'$ is an isomorphism.*

Proposition II.16 *Every surface dominates a minimal surface.*

Proof Let S be a surface. If S is not minimal, there is a birational morphism $S \to S_1$ that is not an isomorphism. If S_1 is not minimal, then there is a birational morphism $S_1 \to S_2$, and so on. Since $\operatorname{rk} NS(S) > \operatorname{rk} NS(S_1) > \operatorname{rk} NS(S_2) > \cdots$ (Remark II.13.1) we eventually get a minimal surface that is dominated by S.

The elements of $B(S)$ are thus obtained by successive blow-ups of minimal surfaces. We shall see later that apart from the ruled surfaces (those birational to a product $C \times \mathbb{P}^1$ for some curve C), every surface has a unique minimal model. The classification problem for surfaces thus falls into two parts; on one hand there are the ruled surfaces, which are known birationally but whose minimal models we have yet to find, and on the other we have the non-ruled surfaces, for which the 'biregular' classification is essentially the same as the birational classification. For these it is therefore enough to classify minimal surfaces.

We shall say that a curve $E \subset S$ is exceptional if it is the exceptional curve of a blow-up $\epsilon : S \to S'$ (S' a smooth surface); hence an exceptional curve E is isomorphic to $\mathbb{P}^1$ and satisfies $E^2 = -1$. By Theorem II.11 a surface is minimal if and only if it contains no exceptional curve.

We close this chapter with a harder theorem than the preceding ones, which gives a numerical characterization of exceptional curves.

Theorem II.17 (Castelnuovo's contractibility criterion) *Let S be a surface and $E \subset S$ a curve isomorphic to $\mathbb{P}^1$ with $E^2 = -1$. Then E is an exceptional curve on S.*

Proof The idea is to modify a hyperplane section of S so that the morphism to $\mathbb{P}^n$ it defines remains an embedding outside E, while E is contracted to a point. The most delicate point is then to check that the image surface is smooth.

(a) Let H be a hyperplane section of S such that $H^1(S, \mathcal{O}_S(H)) = 0$ (such an H exists, since by a general theorem of Serre [FAC], for any hyperplane section H_0 we have $H^1(S, \mathcal{O}_S(nH_0)) = 0$ for all sufficiently large n). Set $k = H.E$ and $H' = H + kE$. Then $\mathcal{O}_S(H)_{|E} \cong \mathcal{O}_E(k)$, $\mathcal{O}_S(E)_{|E} \cong \mathcal{O}_E(-1)$ and $\mathcal{O}_S(H')_{|E} \cong \mathcal{O}_E$, since invertible sheaves on E are determined by their degree. Choose a section s of $\mathcal{O}_S(E)$ whose divisor of zeros is E. For $1 \leqslant i \leqslant k$, consider the exact sequences

$$0 \to \mathcal{O}_S(H + (i-1)E) \to \mathcal{O}_S(H + iE) \to \mathcal{O}_E(k-i) \to 0 .$$

Since $H^1(E, \mathcal{O}_E(r)) = 0$ for $r \geqslant 0$, we get long exact sequences

$$\begin{array}{lll} 0 & \to & H^0(S, \mathcal{O}_S(H+i-1)E) & \to & H^0(S, \mathcal{O}_S(H+iE)) \\ & \xrightarrow{r_i} & H^0(E, \mathcal{O}_E(k-i)) & \to & H^1(S, \mathcal{O}_S(H+(i-1)E)) \\ & \to & H^1(S, \mathcal{O}_S(H+iE)) & \to & 0 , \end{array}$$

for $1 \leqslant i \leqslant k$. We see by induction on i that $H^1(S, \mathcal{O}_S(H+iE)) = 0$ for $1 \leqslant i \leqslant k$ and that the restriction map r_i is surjective.

Choose a basis $s_0, \ldots, s_n$ of $H^0(S, \mathcal{O}_S(H))$ and, for $1 \leqslant i \leqslant k$, elements $a_{i,0}, \ldots, a_{i,k-i}$ of $H^0(S, \mathcal{O}_S(H+iE))$ whose restrictions to E form a basis of $H^0(E, \mathcal{O}_E(k-i))$. Then

$$\{s^k s_0, \ldots, s^k s_n, s^{k-1} a_{1,0}, \ldots, s^{k-1} a_{1,k-1}, \ldots, s a_{k-1,1}, a_{k,0}\}$$

is a basis of $H^0(S, \mathcal{O}_S(H'))$. Let $\phi : S \dashrightarrow \mathbb{P}^n$ denote the rational map defined by the corresponding linear system. Since the map defined by $\{s_0, \ldots, s_n\}$ is an embedding, the restriction of ϕ to $S - E$ is an embedding. Since $a_{k,0}$ induces a non-zero constant function on E, we see that ϕ is defined everywhere and contracts E to the point $p = (0, \ldots, 0, 1)$.

Let S' denote the (possibly singular) surface $\phi(S)$; we thus have a morphism $\epsilon : S \to S'$ which is an isomorphism away from E and contracts E to the point $p \in S'$. By Theorem II.11 it is enough to show that S' is smooth at p.

(b) Let $U \subset S$ be the open neighborhood of E defined by $a_{k,0} \neq 0$. Define sections x, y of $\mathcal{O}_U(-E)$ by $x = a_{k-1,0}/a_{k,0}$, $y = a_{k-1,1}/a_{k,0}$; their restrictions to E form a basis of $H^0(E, \mathcal{O}_E(1))$. Hence, shrinking U if necessary, we can suppose that at no point of U do x, y both vanish. Thus they define a morphism $h_2 : U \to \mathbb{P}^1$. The functions sx, sy on U define a morphism $h_1 : U \to \mathbb{A}^2$; then $(h_1, h_2) : U \to \mathbb{A}^2 \times \mathbb{P}^1$ factors through $\hat{\mathbb{A}}^2$, the blow-up of $\mathbb{A}^2$ at the origin (considered as a subvariety of $\mathbb{A}^2 \times \mathbb{P}^1$; see II.1). The morphism $h : U \to \hat{\mathbb{A}}^2$ has the following properties:

(i) h induces an isomorphism from E to the exceptional curve on $\hat{\mathbb{A}}^2$;

(ii) for all $q \in E$, h is étale in a neighbourhood of q.

(i) follows from the choice of x, y. To prove (ii), we must show that the inverse image under h of a system of local coordinates at $h(q)$ is a system of local coordinates at q. Let u, v (resp. U, V) denote the natural coordinates on $\mathbb{A}^2$ (resp. $\mathbb{P}^1$); recall that $\hat{\mathbb{A}}^2$ is defined in $\mathbb{A}^2 \times \mathbb{P}^1$ by the equation $uV - vU = 0$. We can suppose that $x(q) = 0$ and $y(q) = 1$, so that $h(q)$ has coordinates $u = v = U = 0$, $V = 1$; take v and U/V as local coordinates at $h(q)$. We have $h^*v = sy$ and $h^*(U/V) = x/y$; the first vanishes on E to order 1, while the second (restricted to E) is a local coordinate on E at q. Thus we have a local coordinate system on S at q, which proves (ii).

(c) Finally, we shall use the classical topology of analytic varieties and show that with respect to this there is a neighbourhood U of E such that h induces an isomorphism from U to a neighbourhood V of the exceptional curve of $\hat{\mathbb{A}}^2$. First note that this will prove the theorem. By construction, the functions sx, sy on U come from functions on $\epsilon(U)$; in other words, there is a commutative diagram

$$\begin{array}{ccc} U & \xrightarrow{h} & \hat{\mathbb{A}}^2 \\ {\scriptstyle \epsilon}\downarrow & & \downarrow{\scriptstyle \eta} \\ \epsilon(U) & \xrightarrow{\overline{h}} & \mathbb{A}^2 \end{array}$$

where η is the blow-up of $\mathbb{A}^2$ at the origin and $\overline{h}$ is a morphism ($\epsilon(U)$ is open by the properness of ϵ). If h is an isomorphism from U to V, then $\overline{h}$ is an isomorphism form $\epsilon(U)$ to $\epsilon(V)$; for $\epsilon \circ h^{-1}$ factors through a morphism $\eta(V) \to \epsilon(U)$ (Remark II.13 (2)), which is inverse to $\overline{h}$. Since $\eta(V)$ is open in $\mathbb{A}^2$, $\epsilon(U)$ is smooth; i.e. S' is smooth at p. So the theorem will follow from the next result:

Lemma II.18 *Let $f : X \to Y$ be a continuous map of Hausdorff spaces and $K \subset X$ a compact subset. Suppose that*

(i) *$f_{|K}$ is a homeomorphism;*

(ii) *for all $k \in K$, f is a local homeomorphism near k.*

Then there is a neighbourhood U in X of K and an open subset V of Y such that f induces a homeomorphism $U \xrightarrow{\sim} V$.

Sketch of Proof Shrinking X as necessary, we can assume that f is a local homeomorphism at every point of X.

(a) Set $\Omega = \{(x, y) \in X \times X \mid f(x) \neq f(y) \text{ or } x = y\}$. The fact that f is a local homeomorphism means that Ω is open in $X \times X$, and by hypothesis (i), $K \times K \subset \Omega$.

(b) Let K be a compact subset of a space X and Ω a neighbourhood in $X \times X$ of $K \times K$. Then there is a neighbourhood U in X of K such that $U \times U \subset \Omega$. Indeed, for all $k \in K$ one constructs by compactness an open set $U_k \times V_k$ with $\{k\} \times K \subset U_k \times V_k \subset \Omega$. Then a finite number of these sets, say for $k = k_1, \ldots, k_n$, cover $K \times K$, and the open set $U = (\bigcup_i U_{k_i}) \cap (\bigcap_i V_{k_i})$ does the job.

Now (a) and (b) together imply that $f_{|U}$ is injective, and so is a homeomorphism.

Historical Note II.19

The birational point of view was introduced into the geometry of surfaces in the articles already quoted of Noether, and then those of Enriques ([N1], [E1]). Noether used a definition of exceptional curve ('ausgezeichnet') slightly different from ours; his idea was modified by Enriques, who introduced the classical terminology (from which we have deviated); exceptional curves are said to be 'of the first kind', and 'exceptional curves of the second kind' are those contracted by a rational map (see Exercise II.3). These latter only exist on ruled surfaces; they play a very different role from that of exceptional curves, which is why we have not considered them here.

In 1901 Castelnuovo and Enriques ([C-E]) proved the contractibility criterion known as Castelnuovo's (their proof is more or less that given here) and at the same time proved that non-ruled surfaces have a minimal model, obtained by contracting a finite number of exceptional curves. Although blow-ups appear in their construction, they are not given prominence; the structure theorem for birational maps seems only to have been observed much later by Zariski ([Z1, p. 538], 1944). Zariski then gave a complete clarification of the theory of birational maps and minimal models (see e.g. [Z2]); we have followed his ideas, as they are described in [Sh1].

Exercises II.20

(1) Let C be an irreducible curve on a surface S. Show that there is a morphism $\hat{S} \to S$ consisting of a finite number of blow-ups such that the strict transform of C in $\hat{S}$ is smooth. (Show that blowing up a singular point of C strictly decreases its arithmetic genus.)

(2) Let C be an irreducible curve on S, $p \in C$ and $\hat{C}$ the strict transform of C on the blow-up with centre p. The proximate points of p on C (or infinitely near points of order 1) are by definition the points of $\hat{C}$ lying over p; their multiplicity is by definition their multiplicity on $\hat{C}$. The infinitely near points of order n are the proximate points of the infinitely near points of order $(n-1)$.

(a) Show that the multiplicity m of C at p satisfies

$$m = \hat{C}.E = \sum m_x(\hat{C} \cap E) \geqslant \sum m_x(\hat{C}) ,$$

where the sum runs through the proximate points $x \in \hat{C} \cap E$; find an example with strict inequality.

(b) If C, C' are distinct irreducible curves, show that

$$m_p(C \cap C') = \sum_x m_x(C).m_x(C') ,$$

where x runs over the infinitely near points of p on C and C'.

(c) Let N denote the normalization of C. Show that

$$g(C) = g(N) + \sum_i \frac{1}{2} m_i(m_i - 1) ,$$

where m_i are the multiplicities of the points of C – including the infinitely near points.

(3) We say that an irreducible curve C on S is of the second kind if there is a birational map $\phi : S \dashrightarrow S'$ to a smooth surface S' such that $\phi(C)$ is a point, and ϕ is not defined on the whole of C.

(a) Show that C is a (possibly singular) rational curve, and that with the notation of (2)(c) we have

$$C^2 = \sum m_i^2 - 1 + n, \quad C.K = -\sum m_i - 1 - n \,,$$

where $n \geqslant 0$ and $n > 0$ if C is smooth.

(b) Let C be a rational curve on S with $C^2 \geqslant \sum m_i^2 - 1$ (and also $C^2 \geqslant 0$ if C is smooth). Show that C is a curve of the second kind.

III
RULED SURFACES

Definition III.1 *A surface S is ruled if it is birationally equivalent to $C \times \mathbb{P}^1$, where C is a smooth curve. If $C = \mathbb{P}^1$, S is said to be rational.*

Examples III.2

(a) $C \times \mathbb{P}^1$ is a ruled surface.

(b) More generally, let E be a rank 2 vector bundle over the curve C. Consider the projective bundle $\mathbb{P}_C(E)$ associated with E; it is a surface fibred over C, such that the fibre over a point $x \in C$ is the projective space associated to the vector space E_x. Since E is locally trivial, $\mathbb{P}_C(E)$ is isomorphic (locally over C) to $C \times \mathbb{P}^1$, so it is a ruled surface.

(c) In Chapter IV we will see numerous examples of rational surfaces.

In order to determine the minimal models of ruled surfaces, we will need an auxiliary notion:

Definition III.3 *Let C be a smooth curve. A geometrically ruled surface over C is a surface S, together with a smooth morphism $p : S \to C$ whose fibres are isomorphic to $\mathbb{P}^1$.*

Examples III.2(a) and (b) are geometrically ruled surfaces. It is not obvious a priori that a geometrically ruled surface is ruled. This will follow from the next theorem:

Theorem III.4 (Noether–Enriques) *Let S be a surface, p a morphism from S to a smooth curve C. Suppose there exists $x \in C$ such that p is smooth over x and the fibre $p^{-1}(x)$ is isomorphic to $\mathbb{P}^1$. Then there exists a Zariski open subset U of C containing x and an isomorphism from $p^{-1}(U)$ to $U \times \mathbb{P}^1$ such that the diagram*

$$\begin{array}{ccc} p^{-1}(U) & \xrightarrow{\sim} & U \times \mathbb{P}^1 \\ & {\scriptstyle p}\searrow \quad \swarrow{\scriptstyle pr_1} & \\ & U & \end{array}$$

is commutative. In particular S is ruled.

We begin with an obvious remark, which will often be useful:

Useful Remark III.5 *Let D be an effective divisor, C an irreducible curve such that $C^2 \geqslant 0$. Then $D.C \geqslant 0$.*

Proof Write $D = D' + nC$, where D' does not contain C, and $n \geqslant 0$; then $D.C = D'.C + nC^2 \geqslant 0$.

Proof of Theorem III.4
Step 1: $H^2(S, \mathcal{O}_S) = 0$.

Put $F = p^{-1}(x)$. We have $F^2 = 0$ and $F.K = -2$ (I.8(i) and the genus formula). Suppose $H^2(S, \mathcal{O}_S) \neq 0$; then $|K|$ contains an effective divisor D (by Serre duality, I.11). We have $D.F = -2$, but also $D.F \geqslant 0$ by the useful remark, a contradiction.

Step 2: There exists a divisor H of S such that $H.F = 1$.

Let f be the class of F in $H^2(S, \mathbb{Z})$. Since $H^2(S, \mathcal{O}_S) = 0$, the map $\text{Pic}(S) \to H^2(S, \mathbb{Z})$ is surjective (I.10). Thus it suffices to show that there exists a class $h \in H^2(S, \mathbb{Z})$ with $h.f = 1$. As a runs through $H^2(S, \mathbb{Z})$, the set of integers $(a.f)$ is an ideal in $\mathbb{Z}$, of the form $d\mathbb{Z}$ $(d \geqslant 1)$. The map $a \mapsto \frac{1}{d}(a.f)$ is a linear form on $H^2(S, \mathbb{Z})$; now Poincaré duality says that the cup product $H^2(S, \mathbb{Z}) \otimes H^2(S, \mathbb{Z}) \to H^4(S, \mathbb{Z}) \xrightarrow{\sim} \mathbb{Z}$ is a duality, in other words that the associated map

$$H^2(S, \mathbb{Z}) \to \text{Hom}(H^2(S, \mathbb{Z}), \mathbb{Z})$$

is surjective, with kernel equal to the torsion subgroup. Hence there exists an element $f' \in H^2(S, \mathbb{Z})$ such that

$$(a.f') = \frac{1}{d}(a.f) \quad \text{for all} \quad a \in H^2(S, \mathbb{Z}) ,$$

so that $f = df'$ modulo torsion in $H^2(S, \mathbb{Z})$.

Notice that if k is the class of K in $H^2(S, \mathbb{Z})$, the integer $a^2 + a.k$ is even for every divisor class a (hence in this case for all $a \in H^2(S, \mathbb{Z})$); for the expression is linear in a (mod 2), and is even for irreducible a by the genus formula. Since $f^2 = 0$ and $f.k = -2$ we have $f'^2 = 0$ and

Step 3: Consider the exact sequence

$$0 \to \mathcal{O}_S(H+(r-1)F) \to \mathcal{O}_S(H+rF) \to \mathcal{O}_F(1) \to 0 \qquad (r \in \mathbb{Z}).$$

It gives the long exact cohomology sequence

$$H^0(S,\mathcal{O}_S(H+rF)) \xrightarrow{a_r} H^0(F,\mathcal{O}_F(1)) \to H^1(S,\mathcal{O}_S(H+(r-1)F)) \xrightarrow{b_r} H^1(S,\mathcal{O}_S(H+rF)) \to 0\,.$$

The sequence of quotient spaces $H^1(S,\mathcal{O}_S(H+rF))$ must become stationary for r sufficiently large; so there exists r such that b_r is bijective, and a_r is surjective. Let V be a vector subspace of $H^0(S,\mathcal{O}_S(H+rF))$ of dimension 2 such that $a_r(V) = H^0(F,\mathcal{O}_F(1))$; let P be the corresponding pencil. It may have fixed components, but they must be contained in fibres $F_{x_1},\ldots,F_{x_k}$ of p distinct from F (since P has no base points on F). Similarly any base points of the mobile part of P are contained in fibres $F_{x_{k+1}},\ldots,F_{x_l}$ distinct from F. Denote by $F_{x_{l+1}},\ldots,F_{x_m}$ those fibres of p which are reducible. Put $U = C - \{x_1,\ldots,x_m\}$. The pencil P', i.e. P restricted to U, is base-point free. Every curve C_t in P' is the union of a section of p and possibly a number of fibres; but in fact C_t contains no fibres, otherwise it would meet the curve $C_{t'}$ $(t \neq t')$, and P' would have base points. Hence the divisors of P' are all sections $(C_t)_{t\in\mathbb{P}^1}$ of the fibration p. Since the pencil P' is without base points, it defines a morphism $g : p^{-1}(U) \to \mathbb{P}^1$, with fibre $g^{-1}(t) = C_t$. Consider the morphism $h = (p,g)$ from $p^{-1}(U)$ to $U \times \mathbb{P}^1$. Since $h^{-1}((y,t)) = F_y \cap C_t$, h is an isomorphism, which completes the proof of the theorem.

Remark III.6 The following is a more highbrow proof of Step 3: Set $E = p_*(\mathcal{O}_S(H))$; let U be an open subset of C over which p is smooth. It follows from general theorems on proper morphisms that $E_{|U}$ is locally free of rank 2, and that the natural map $p^*E \to \mathcal{O}_S(H)$ is surjective over $p^{-1}(U)$. We obtain a U-morphism $p^{-1}(U) \to \mathbb{P}_U(E)$ which is an isomorphism fibre by fibre, and the result follows. This stage is purely formal; the important part of the proof is Step 2, which is usually proved by quoting Tsen's theorem.

Proposition III.7 *Every geometrically ruled surface over C is C-isomorphic to $\mathbb{P}_C(E)$ for some rank 2 vector bundle over C. The bundles $\mathbb{P}_C(E)$, $\mathbb{P}_C(E')$ are C-isomorphic if and only if there exists a line bundle L on C such that $E' \cong E \otimes L$.*

Proof The Noether–Enriques theorem shows that the fibration $p : S \to C$ is locally trivial (i.e. there is a cover (U_μ) of C and U_μ-isomorphisms $t_\mu : p^{-1}(U_\mu) \xrightarrow{\sim} U_\mu \times \mathbb{P}^1$). By well-known formalism (see for example [H, Chapter I, 3.2]), the set of isomorphism classes of such bundles can be identified with the cohomology set $H^1(C, G)$, where G is the sheaf of (non-commutative) groups defined by

$$\begin{aligned} G(U) &= \mathrm{Aut}_U(U \times \mathbb{P}^1) \\ &= \{\text{morphisms of } U \text{ into the algebraic group } PLG(2, \mathbb{C})\} \end{aligned}$$

(recall the principle of this identification: by composing the trivialisations t_μ and t_λ^{-1} restricted to $p^{-1}(U_\mu \cap U_\lambda)$, we obtain a Čech 1-cocycle $(\mu, \lambda) \mapsto g_{\mu\lambda} \in G(U_\mu \cap U_\lambda)$, and hence a class in $H^1(C, G)$).

Write $G = PGL(2, \mathcal{O}_C)$. From the exact sequence

$$1 \to \mathbb{C}^* \to GL(2, \mathbb{C}) \to PGL(2, \mathbb{C}) \to 1$$

we obtain an exact sequence of sheaves

$$1 \to \mathcal{O}_C^* \to GL(2, \mathcal{O}_C) \to PGL(2, \mathcal{O}_C) \to 1 ,$$

where $GL(2, \mathcal{O}_C)$ is the sheaf of 2×2 invertible matrices with coefficients in $\mathcal{O}_C$. We then get an exact sequence of cohomology sets

$$H^1(C, \mathcal{O}_C^*) \to H^1(C, GL(2, \mathcal{O}_C)) \to H^1(C, PGL(2, \mathcal{O}_C)) \to H^2(C, \mathcal{O}_C^*)$$

where $H^1(C, GL(2, \mathcal{O}_C))$ is the set of isomorphism classes of rank 2 vector bundles on C; it is easy to see that the group $H^1(C, \mathcal{O}_C^*) \cong \mathrm{Pic}\, C$ acts on this set by tensor product. The last assertion of the proposition follows from the exact sequence; so does the first if we know that $H^2(C, \mathcal{O}_C^*) = 0$. Now there are two good reasons (at least) for this: the first is that since $\dim C = 1$, we have $H^2(C, F) = 0$ for all sheaves on C (with the Zariski topology; cf. [FAC]); the second is that anyway $H^2(T, \mathcal{O}_T^*) = 0$ for every smooth variety T, because of the exact sequence

$$1 \to \mathcal{O}_T^* \to \mathcal{K}_T^* \to \mathrm{Div}(T) \to 0$$

which defines a 2-step resolution of $\mathcal{O}_T^*$ by flasque sheaves.

Lemma III.8 *Let S be a minimal surface, C a smooth curve, $p : S \to C$ a morphism with generic fibre isomorphic to $\mathbb{P}^1$. Then S is geometrically ruled by p (i.e. C-isomorphic to a projective bundle $\mathbb{P}_C(E)$).*

Proof Let F be a fibre of p; then $F^2 = 0$, $F.K = -2$. First assume that F is irreducible. The argument (III.4, Step 2) shows that F cannot be multiple; the genus formula (and Remark I.16(1)) then shows that F is isomorphic to $\mathbb{P}^1$, and that p is smooth over F. Thus it suffices to show that F cannot be reducible.

Lemma III.9 *Let p be a surjective morphism from a surface to a smooth curve with connected fibres, $F = \sum n_i C_i$ a reducible fibre of p. Then $C_i^2 < 0$ for all i.*

Proof We have $n_i C_i^2 = C_i.(F - \sum_{j \neq i} n_j C_j)$. Now $C_i F = 0$ since we can replace F by another fibre (cf. I.8(i)), $(C_i.C_j) \geqslant 0$ for $i \neq j$, and $(C_i.C_j) > 0$ for at least one j because F is connected.

Conclusion of Proof of III.8 Let $F = \sum n_i C_i$ be a reducible fibre; the genus formula and Lemma III.9 show that $K.C_i \geqslant -1$, with equality if and only if $C_i^2 = -1$, $g(C_i) = 0$; these two conditions imply that C_i is an exceptional curve, which is impossible under our hypothesis. Thus $K.C_i \geqslant 0$ for all i, and $K.F \geqslant 0$, which contradicts $K.F = -2$.

Theorem III.10 *Let C be a smooth irrational curve. The minimal models of $C \times \mathbb{P}^1$ are the geometrically ruled surfaces over C, that is the projective bundles $\mathbb{P}_C(E)$.*

Proof We will show that a geometrically ruled surface $p : S \to C$ contains no exceptional curves. Such a curve E cannot be a fibre of p since $E^2 = -1$; so we must have $p(E) = C$, implying C is rational, which contradicts the hypothesis.

Now let S be a minimal surface, ϕ a birational map from S to $C \times \mathbb{P}^1$, and q the projection from $C \times \mathbb{P}^1$ to C. Consider the rational map $q \circ \phi : S \dashrightarrow C$; by the theorem on elimination of indeterminacy (II.7), there exists a commutative diagram

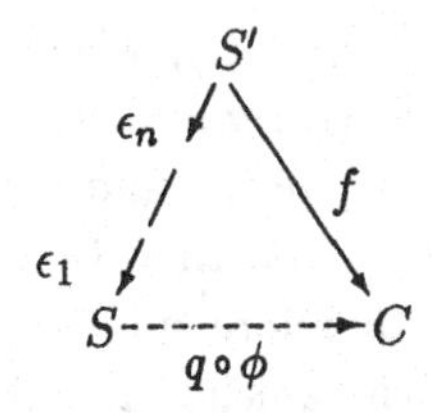

where the ϵ_i are blow-ups, and f is a morphism. We can assume that n is the minimal number of blow-ups necessary for a diagram of this type to exist. Suppose $n > 0$, and let E be the exceptional curve of the blow-up ϵ_n; since C is not rational, $f(E)$ is necessarily reduced to a point, so f factorizes as $f' \circ \epsilon_n$ (Remark II.13.(2)), contradicting the minimality of n. It follows that $n = 0$, and $q \circ \phi$ is a morphism with generic fibre isomorphic to $\mathbb{P}^1$. Lemma III.8 completes the proof.

Notice that this proof cannot be applied to rational surfaces. We postpone the search for minimal rational surfaces until Chapter V, and try for now to summarize the classification of geometrically ruled surfaces over a given curve C. By Proposition III.7, this is the same as classifying the rank 2 vector bundles on C, up to tensor product with a line bundle. The theory of vector bundles on a curve is delicate, but can be taken as well understood – cf. [R]. We will content ourselves with some elementary remarks showing that there are 'lots' of minimal ruled surfaces.

Let C be a smooth curve, E a rank 2 vector bundle on C. We identify E with its sheaf of local sections, which is locally free of rank 2. Write $\deg(E) = \deg(\wedge^2 E)$, $h^i(E) = \dim H^i(C, E)$, $\chi(E) = h^0(E) - h^1(E)$.

Observe that $\deg(E \otimes L) = \deg(E) + 2\deg(L)$ for L in $\operatorname{Pic} C$, which allows us to alter $\deg(E)$ to any value of the same parity.

Lemma III.11

(i) *There exists an exact sequence* $0 \to L \to E \to M \to 0$ *with* L, M *in* $\operatorname{Pic} C$.

(ii) *If* $h^0(E) \geqslant 1$, *we can take* $L = \mathcal{O}_C(D)$, *with* $D \geqslant 0$.

(iii) *If* $h^0(E) \geqslant 2$ *and* $\deg(E) > 0$, *we can assume* $D > 0$.

Proof Note that we can assume $h^0(E) \geqslant 1$ in proving (i): we can replace E by $E \otimes N$, and for suitable $N \in \operatorname{Pic} S$ this bundle admits sections.

Assume then that E admits a non-zero section s; we obtain from it a non-zero morphism $\check{s}$ from the dual $\check{E}$ of E to $\mathcal{O}_C$. Its image is an ideal of $\mathcal{O}_C$, that is a sheaf $\mathcal{O}_C(-D)$, where D is an effective divisor of C. The kernel of the surjective morphism $\check{E} \to \mathcal{O}_C(-D)$ is thus an invertible sheaf. Taking duals gives (i) and (ii).

It is clear from this construction that D is the divisor of zeros of a section s. To prove (iii), it therefore suffices to show that there exists a non-zero section of E which vanishes at some point. Let s, t be two linearly independent sections; the section $s \wedge t$ of $\wedge^2 E$ must vanish at some point p of C (since $\deg(E) > 0$), which means that there exist μ, λ (not both zero) such that $\mu s(p) + \lambda t(p) = 0$. So the section $\mu s + \lambda t$ vanishes at p, hence the result.

Corollary III.12 (Riemann–Roch for rank 2 vector bundles) *We have*

$$\chi(E) = \deg(E) + 2 - 2g(C) .$$

Proof By III.11(i) and Riemann–Roch for invertible sheaves, we have:

$$\begin{aligned} \chi(E) &= \chi(L) + \chi(M) = \deg(L) + \deg(M) + 2(1 - g(C)) \\ &= \deg(E) + 2 - 2g(C). \end{aligned}$$

Remark III.13 The geometrical interpretation of III.11(i) is that the fibration $p : \mathbb{P}_C(E) \to C$ admits a section (compare III.17 later on). This can be shown directly: the fibration p, being locally trivial, admits a local section, that is a rational map $s : C \dashrightarrow \mathbb{P}_C(E)$ such that $p \circ s = \mathrm{Id}_C$. But every rational map from C to a complete variety is a morphism, hence the result.

III.14 Extension of sheaves

Given the extension of M by L in III.11 (i), it is natural to ask if it is trivial, in other words if $E \cong L \oplus M$. Equivalently, does the exact sequence

$$0 \to L \otimes M^{-1} \to E \otimes M^{-1} \to \mathcal{O}_C \to 0$$

split?; but this is the case if and only if it has a section, that is if there exists a section of $H^0(C, E \otimes M^{-1})$ which maps onto $1 \in H^0(C, \mathcal{O}_C)$. Using the exact cohomology sequence

$$H^0(C, E \oplus M^{-1}) \longrightarrow H^0(C, \mathcal{O}_C) \xrightarrow{\partial} H^1(C, L \otimes M^{-1})$$

that would mean $\partial(1) = 0$. The class $\partial(1) \in H^1(C, L \otimes M^{-1})$ is called the class of the extension $L \to E \to M$; its vanishing is necessary and sufficient for the extension to be trivial.

More generally, it is easy to see that two extensions are isomorphic if their classes are proportional. In particular if $h^1(L \otimes M^{-1}) = 1$, there is – up to isomorphism – only one non-trivial extension of M by L.

Proposition III.15

(i) *Every rank 2 vector bundle on $\mathbb{P}^1$ is decomposable, i.e. the sum of two invertible sheaves. In particular every geometrically ruled surface over $\mathbb{P}^1$ is isomorphic to one of the surfaces*

$$\mathbb{F}_n = \mathbb{P}_{\mathbb{P}^1}(\mathcal{O}_{\mathbb{P}^1} \oplus \mathcal{O}_{\mathbb{P}^1}(n)) \quad \textit{for} \;\; n \geqslant 0.$$

(ii) *Every rank 2 vector bundle on an elliptic curve is either decomposable, or isomorphic to $E \otimes L$, where $L \in \operatorname{Pic} C$ and E is one of the following bundles:*
either *the non-trivial extension of $\mathcal{O}_C$ by $\mathcal{O}_C$;*
or *for any $p \in E$, the non-trivial extension of $\mathcal{O}_C(p)$ by $\mathcal{O}_C$.*

(iii) *For every curve C of genus g, there exist (open) varieties S and families of rank 2 vector bundles parametrized by S : $(E_s)_{s \in S}$, such that $\wedge^2 E_s$ is constant, E_s is indecomposable, and $E_s \not\cong E_{s'}$ for $s \neq s'$; such families can be found with $\dim S \geqslant 2g - 3$.*

(The right dimension is in fact $3g - 3$; we simply wish to show that there are 'more' indecomposable than decomposable bundles.)

Proof

(i) Let E be a rank 2 vector bundle on $\mathbb{P}^1$; by replacing E by $E \otimes L$, we can assume that $d = \deg(E) = 0$ or -1. By Riemann–Roch we then have $h^0(E) \geqslant d+2 \geqslant 1$, so there exists an exact sequence

$$0 \to \mathcal{O}_{\mathbb{P}^1}(k) \to E \to \mathcal{O}_{\mathbb{P}^1}(d-k) \to 0$$

with $k \geqslant 0$ (III.11(ii)). The class of this extension lies in $H^1(\mathbb{P}^1, \mathcal{O}_{\mathbb{P}^1}(2k-d))$ which is always zero, hence (i).

(ii) Let E be a rank 2 vector bundle on an elliptic curve C; we may assume $\deg(E) = 1$ or 2, so that $h^0(E) \geqslant 1$ by Riemann–Roch. Thus there exist invertible sheaves L_k and M_{d-k}, of degrees $k \geqslant 0$ and $d - k$, and an exact sequence:

$$0 \to L_k \to E \to M_{d-k} \to 0.$$

Moreover if $d = 2$ we can assume $k \geqslant 1$ (III.11 (iii)), and if $k = 0$, $L_0 = \mathcal{O}_C$.
The class of the extension lives in $H^1(C, L_k \otimes M_{d-k}^{-1})$, which is zero unless either

$d = 1$, $k = 0$: then E is the extension of a line bundle of degree 1 by $\mathcal{O}_C$;

or

$d = 2$, $k = 1$ and $L \otimes M_1^{-1} \cong \mathcal{O}_C$: then $E \otimes L_1^{-1}$ is the extension of $\mathcal{O}_C$ by $\mathcal{O}_C$.

This proves (ii). We note that the bundle E_p (resp. E_0) which is an extension of $\mathcal{O}_C(p)$ (resp. $\mathcal{O}_C$) by $\mathcal{O}_C$ is indecomposable: this follows at once from the fact that it contains $\mathcal{O}_C$ as a sub-bundle, and $h^0(E_p) = h^0(E_0) = 1$. Moreover, we leave it as an exercise to see that $E_q \cong E_p \otimes L$, where $L^2 = \mathcal{O}_C(q-p)$, so that up to isomorphism there are only 2 geometrically ruled surfaces over C corresponding to indecomposable bundles.

(iii) Pick an invertible sheaf L on C with degree $g-1$ and $h^0(L)=0$. For S we take an affine hyperplane of $H^1(C, L^{-1})$ not passing through the origin. With each $s \in S$ we associate the bundle defined by the extension

$$(\mathcal{E}_s) \qquad 0 \to \mathcal{O}_C \to E_s \to L \to 0$$

of class s. Then $(E_s)_{s\in S}$ forms a family of bundles over C parameterized by S. Since $h^0(E_s)=1$, the extension $\mathcal{E}_s$ is uniquely determined by the bundle E_s: it follows that $E_s \not\cong E_{s'}$, for $s \neq s'$, and that E_s is indecomposable.

Remark III.16 We have studied geometrically ruled surfaces up to C-isomorphism; it is also natural to consider the situation up to isomorphism. In Chapter IV we will look at rational surfaces; here we assume $C \not\cong \mathbb{P}^1$. Then for every isomorphism $v : \mathbb{P}_C(E) \xrightarrow{\sim} \mathbb{P}_C(E')$, there is an automorphism u of C and a commutative diagram

$$\begin{array}{ccc} \mathbb{P}_C(E) & \xrightarrow{v} & \mathbb{P}_C(E') \\ {\scriptstyle p}\downarrow & & \downarrow{\scriptstyle p'} \\ C & \xrightarrow{u} & C\,. \end{array}$$

In fact we take $u = p'vs$, where s is an arbitrary section of p. Since C is not rational, v maps fibres of p to fibres of p'; it follows easily that $up = p'v$.

Thus the surfaces $\mathbb{P}_C(E)$ and $\mathbb{P}_C(E')$ are isomorphic if and only if there exists an automorphism u of C, and a line bundle L on C, such that $u^*E' \cong E \otimes L$.

We are now going to describe the Picard group of geometrically ruled surfaces.

III.17 Projective bundles Let $S = \mathbb{P}_C(E)$, $p : S \to C$ its structural morphism. The bundle p^*E on S has a line bundle N as a sub-bundle in a natural way: above a point $s \in S$, corresponding to a line $D \subset E_{p(s)}$, we have $N_s = D$. The bundle $\mathcal{O}_S(1)$, often called the tautological bundle of S, is defined by the exact sequence

$$0 \to N \to p^*E \xrightarrow{u} \mathcal{O}_S(1) \to 0\,.$$

Let T be a variety, $f : T \to C$ a morphism. To each C-morphism $g : T \to \mathbb{P}_C(E)$ we associate the line bundle $L = g^*\mathcal{O}_S(1)$ and the surjective morphism $g^*u : f^*E \to L$. Conversely, given a line bundle L on T and a surjective morphism $v : f^*E \to L$, we define a C-morphism $g : T \to \mathbb{P}_C(E)$ by associating to a point t in T the line $\operatorname{Ker}(v_t) \subset E_{f(t)}$. These two constructions are inverse to each other.

In particular, giving a section $s : C \to S$ of p is equivalent to giving a quotient line bundle of E.

Proposition III.18 *Let $S = \mathbb{P}_C(E)$ be a geometrically ruled surface over C, $p : S \to C$ the structure map. Write h for the class of the sheaf $\mathcal{O}_S(1)$ in $\operatorname{Pic} S$ (or in $H^2(S, \mathbb{Z})$). Then*

(i) $\operatorname{Pic} S = p^* \operatorname{Pic} C \oplus \mathbb{Z}h$.

(ii) $H^2(S, \mathbb{Z}) = \mathbb{Z}h \oplus \mathbb{Z}f$, *where f is the class of a fibre.*

(iii) $h^2 = \deg(E)$.

(iv) $[K] = -2h + (\deg(E) + 2g(C) - 2)f$ *in* $H^2(S, \mathbb{Z})$.

Proof First we prove (i). Let F be a fibre of p; since $F.h = 1$, every element of $\operatorname{Pic} S$ can be written $D + mh$, where $D.F = 0$; so it is enough to prove that D is the pull-back of a divisor on C. Put $D_n = D + nF$; then $D_n^2 = D^2$, $D_n.K = D.K - 2n$, and $h^0(K - D_n) = 0$ whenever n is sufficiently large. The Riemann–Roch theorem then gives that $h^0(D_n) \geqslant n + \text{const.}$, so that the system $|D_n|$ is non-empty for sufficiently large n. Let $E \in |D_n|$; since $E.F = 0$, every component of E is a fibre of p, so E is the inverse image by p of a divisor on C.

Part (ii) can be proved either directly from the topology (using exact sequences of S^2-bundles), or from part (i): since $H^2(S, \mathbb{Z})$ is a quotient of $\operatorname{Pic} S$ (cf. Theorem III.4, Step 1) and two points of C have the same cohomology class in $H^2(C, \mathbb{Z}) \cong \mathbb{Z}$, $H^2(S, \mathbb{Z})$ is generated by the elements f and h; these are linearly independent since $f^2 = 0$, $f.h = 1$.

To prove (iii), we will make use of the following result: let E' be a vector bundle on a surface S such that there exists an exact sequence

$$0 \to L \to E' \to M \to 0 \quad \text{with} \quad L, M \in \operatorname{Pic} S\ ;$$

then

$$\begin{aligned} L.M &= L^{-1}.M^{-1} \\ &= \chi(\mathcal{O}_S) - \chi(L) - \chi(M) + \chi(L \otimes M) \qquad \text{(by I.4)} \\ &= \chi(\mathcal{O}_S) - \chi(E') + \chi(\wedge^2 E')\ . \end{aligned}$$

In particular, the number $(L.M)$ depends only on E'; we will denote it by $c_2(E')$.

We now apply this to the bundle p^*E on S. Since there is an exact sequence $0 \to L \to E \to M \to 0$, we have $c_2(p^*E) = (p^*L.p^*M) = 0$. On the other hand, the exact sequence

$$0 \to N \to p^*E \to \mathcal{O}_S(1) \to 0$$

gives $h.[N] = 0$. From this exact sequence we obtain an isomorphism $N \otimes \mathcal{O}_S(1) \cong p^* \wedge^2 E$, whence $[N] = -h + p^*e$, writing e for the class of $\wedge^2 E$ in $\operatorname{Pic} C$. It follows that $h^2 = h.p^*e = \deg(E)$.

From (ii) we have $[K] = ah + bf$ in $H^2(S, \mathbb{Z})$, with $a, b \in \mathbb{Z}$; also $a = K.f = -2$. Let $s : C \to S$ be a section of p such that $[s(C)] = h + rf$ in $H^2(S, \mathbb{Z})$ for some integer r. The genus formula for $s(C)$ can be written:

$$2g(C) - 2 = (h + rf)^2 + (h + rf)(-2h + bf) = -\deg(E) + b\ ,$$

which gives (iv).

Numerical Invariants

We end this chapter with a calculation of the numerical invariants of a ruled surface. To every surface S we can associate several integers; for example, the algebro-geometric invariants:

$$\begin{aligned} q(S) &= h^1(S, \mathcal{O}_S) \\ p_g(S) &= h^2(S, \mathcal{O}_S) = h^0(S, \mathcal{O}_S(K)) \text{ (by Serre duality)} \\ P_n(S) &= h^0(S, \mathcal{O}_S(nK)) \text{ for } n \geqslant 1. \end{aligned}$$

The P_n are called the plurigenera of S, $p_g = P_1$ is the geometric genus, and q is the irregularity of S. We have $\chi(\mathcal{O}_S) = 1 - q(S) + p_g(S)$.

We can also consider the topological invariants:

$$b_i(S) = \dim_{\mathbb{R}} H^i(S, \mathbb{R})\ , \quad \chi_{\text{top}}(S) = \sum_i (-1)^i b_i(S)\ .$$

We have $b_0 = b_4 = 1$ and $b_3 = b_1$ by Poincaré duality, so that $\chi_{\text{top}}(S) = 2 - 2b_1(S) + b_2(S)$.

When there is no ambiguity about S we simply write q, p_g, P_n, b_i. These invariants are related by the following equation, which comes from Hodge theory ([W]) and which we state without proof.

Fact III.19 $q(S) = h^0(S, \Omega^1_S) = \frac{1}{2} b_1(S)$.

Recall that we are also assuming Noether's formula (I.14):

$$\chi(\mathcal{O}_S) = \frac{1}{12}(\chi_{\text{top}}(S) + K_S^2)\ .$$

Proposition III.20 *The integers q, p_g, P_n are birational invariants.*

First we show the birational invariance of $q(S) = h^0(S, \Omega_S^1)$. Let $\phi : S' \dashrightarrow S$ be a birational map; it corresponds to a morphism $f : S' - F \to S$, where F is a finite set. For every 1-form $\omega \in H^0(S, \Omega_S^1)$, the form $f^*\omega$ defines a rational 1-form on S', with poles lying in F; since the poles of a differential form are divisors, $f^*\omega$ is in fact holomorphic on the whole of S'. This enables us to define an injective map $\phi^* : H^0(S, \Omega_S^1) \to H^0(S', \Omega_{S'}^1)$; since ϕ is birational, ϕ^* has an inverse, so $q(S) = q(S')$.

The birational invariance of p_g and P_n is proved in the same way.

We remark that K^2 and b_2 are not birational invariants (cf. Proposition II.3).

Proposition III.21 *Let S be a ruled surface over C; then*

$$q(S) = g(C)\ ;\quad p_g(S) = 0\ ;\quad P_n(S) = 0 \quad \textit{for all} \quad n \geqslant 2.$$

If S is geometrically ruled, then

$$K_S^2 = 8(1 - g(C))\ ,\quad b_2(S) = 2\ .$$

In Chapter VI we will see that the condition $P_n = 0$ for all n characterizes the ruled surfaces. To prove III.21, we will use the following general fact:

Fact III.22 *Let X, Y be two varieties, p, q the projections of $X \times Y$ on X, Y.*

(i) *If F (resp. G) is a vector bundle on X (resp. Y), the canonical homomorphism $H^0(X, F) \otimes H^0(Y, G) \to H^0(X \times Y, p^*F \otimes q^*G)$ is an isomorphism.*

(ii) *If X and Y are smooth, then $\Omega_{X\times Y}^1 \cong p^*\Omega_X^1 \oplus q^*\Omega_Y^1$.*

We briefly recall the proof of (i): the projection formula (whose proof is elementary) gives an isomorphism from $p_*(p^*F \otimes q^*G)$ to $F \otimes p_*q^*G$; the same formula applied to the sheaf q^*G on $U \times Y$, for any open subset U of Y, shows that $p_*q^*G \cong \mathcal{O}_X \otimes_{\mathbb{C}} H^0(Y, G)$; (i) follows immediately. Assertion (ii) comes immediately from the following fact: if $x_1, \ldots, x_p$

(resp., $y_1, \ldots, y_q$) is a system of local coordinates for X (resp. Y), then $x_1, \ldots, x_p, y_1, \ldots, y_q$ is a system of local coordinates for $X \times Y$.

Proof of III.21 To calculate the birational invariants of S, we may assume that $S = C \times \mathbb{P}^1$. Then by III.22 we have:

$$H^0(S, \Omega_S^1) \cong H^0(C, \omega_C) \oplus H^0(\mathbb{P}^1, \omega_{\mathbb{P}^1}), \quad \text{hence } q(S) = g(C);$$
$$H^0(S, \omega_S^{\otimes n}) \cong H^0(C, \omega_S^{\otimes n}) \otimes H^0(\mathbb{P}^1, \omega_{\mathbb{P}^1}^{\otimes n}) = 0;$$

hence $P_n(S) = 0$ ($n \geqslant 1$). If S is geometrically ruled, the calculations of K^2 and b_2 follow easily from III.18.

Historical Note III.23

The Noether–Enriques theorem was proved by Noether ([N3]) for rational surfaces, and by Enriques in the general case ([E2]).

Theorem III.10 is due to Severi ([Se1]). The 'arithmetic genus' $p_a = p_g - q$ and the geometric genus p_g were introduced by Clebsch and Noether (cf. [N1]), who demonstrated their birational nature (the terminology comes from the fact that if one takes a representative of the surface in $\mathbb{P}^3$, with a double curve and triple points, the 'arithmetic' or 'numerical' genus is defined by an explicit formula as a function of the degree of the surface, the degree and genus of the double curve, etc.). It seems to have been suspected at first that $p_a = p_g$, but Cayley pointed out that $p_a < p_g$ for irrational ruled surfaces (1871). The surfaces for which this strict inequality held were thought to be exceptional and were termed irregular: the number $q = p_g - p_a$ (positive or zero by definition) was called the irregularity.

Exercises III.24

Let C be a curve, E a rank 2 vector bundle on C, $S = \mathbb{P}_C(E)$.

(1) Let $s \in S$, and F be the fibre of the projection passing through s. Show that on the blow-up of S at s we can contract the strict transform of F. The surface obtained is a geometrically ruled surface S'; the rational map $S \dashrightarrow S'$ is called an elementary transformation with centre s.

(2) A point s of S corresponds to a surjective homomorphism $u_s : E \to \mathbb{C}(s)$, where $\mathbb{C}(s)$ is the sheaf which is zero outside s and has stalk $\mathbb{C}$ at s (compare III.17). Show that $E' = \operatorname{Ker}(u_s)$ is a rank 2 vector bundle on C, that $S' \cong \mathbb{P}_C(E')$ (Exercise 1) and that the elementary transformation $S \dashrightarrow S'$ corresponds to the inclusion $E' \to E$.

(3) Assume $C \not\cong \mathbb{P}^1$; let X be a minimal surface, $\phi : X \dashrightarrow S$ a birational map. Without using Theorem III.10, show that ϕ is composed of an isomorphism and some elementary transformations (if $n(\phi)$ is the minimum number of blow-ups necessary to make ϕ everywhere defined, show that there exists an elementary transformation $t : \mathbb{P}_C(E) \dashrightarrow \mathbb{P}_C(E')$ such that $n(t \circ \phi) < n(\phi)$). This gives another proof of Theorem III.10.

(4) Let $\mathrm{Aut}_b(S)$ be the group of birational automorphisms of S. If $C \not\cong \mathbb{P}^1$, show that there is an exact sequence

$$1 \to PGL(2, K) \to \mathrm{Aut}_b(S) \to \mathrm{Aut}(C) \to 1$$

where K is the rational function field of C. Also, the choice of a birational map from S to $C \times \mathbb{P}^1$ gives rise to a natural splitting of this sequence.

(5) Let $\mathbb{F}_n = \mathbb{P}_{\mathbb{P}^1}(\mathcal{O} \oplus \mathcal{O}(n))$, $s \in \mathbb{F}_n$. Show that the elementary transformation of $\mathbb{F}_n$ with centre s is isomorphic to $\mathbb{F}_{n-1}$ or to $\mathbb{F}_{n+1}$, depending on the position of s. Distinguish the two cases.

(6) Show that the surface $\mathbb{F}_1$ contains an exceptional line B. We also call the following two operations elementary transformations:

(a) blowing up a point of $\mathbb{F}_1$ not lying on B, then contracting B;
(b) the automorphism $(x, y) \mapsto (y, x)$ of $\mathbb{F}_0 = \mathbb{P}^1 \times \mathbb{P}^1$.

(7) Let S be a minimal rational surface, $\phi : S \dashrightarrow \mathbb{F}_n$ a birational map. Show that:
either ϕ is composed of an isomorphism and elementary transformations; in particular $S \cong \mathbb{F}_m$ for $m \neq 1$;
or $n = 1$, $S \cong \mathbb{P}^2$, and ϕ is the inverse of a blow-up.
(Cf. Hartshorne, *Curves with high self-intersection on algebraic surfaces*, Publ. Math. IHES, **36** (1969), 111–125.)

(8) Show that up to homeomorphism there are only two different types of geometrically ruled surfaces over C, corresponding to the parity of $\deg(E)$ (recall that every exact sequence of bundles on C splits, and that two line bundles of the same degree are topologically isomorphic; if L_i is a line bundle of degree i, show that $L_p \oplus L_{2r-p} \cong (L_r)^2$).

(9) Let S be a surface of degree 4 in $\mathbb{P}^3$ containing two non-coplanar double lines. Show that the normalization of S is a geometrically ruled surface over an elliptic curve.

(10) Let $S \subset \mathbb{P}^3$ be a surface, not necessarily smooth, such that every point of S lies on a line contained in S. Show that S is birationally

isomorphic to a ruled surface. Conversely show that every ruled surface is birational to a surface of the preceding type (and more precisely, we can take this to be a cone in $\mathbb{P}^3$).

IV
RATIONAL SURFACES

These are surfaces birational to $\mathbb{P}^2$. We shall begin by studying those surfaces geometrically ruled over $\mathbb{P}^1$, and then give some simple examples of rational surfaces embedded in $\mathbb{P}^n$.

The Surfaces $\mathbb{F}_n$

Recall (III.15) that the only surfaces geometrically ruled over $\mathbb{P}^1$ are the surfaces $\mathbb{F}_n = \mathbb{P}_{\mathbb{P}^1}(\mathcal{O}_{\mathbb{P}^1} \oplus \mathcal{O}_{\mathbb{P}^1}(n))$, $n \geqslant 0$. We denote by h (resp. f) the class in $\operatorname{Pic} \mathbb{F}_n$ of the tautological bundle $\mathcal{O}_{\mathbb{F}_n}(1)$ (resp. of a fibre).

Proposition IV.1

(i) $\operatorname{Pic} \mathbb{F}_n = \mathbb{Z}h \oplus \mathbb{Z}f$, *with* $f^2 = 0$, $f.h = 1$, $h^2 = n$.

(ii) *If* $n > 0$, *then there is a unique irreducible curve* B *on* $\mathbb{F}_n$ *with negative self-intersection. If* b *is its class in* $\operatorname{Pic} \mathbb{F}_n$, *then* $b = h - nf$, $b^2 = -n$.

(iii) $\mathbb{F}_n$ *and* $\mathbb{F}_m$ *are not isomorphic unless* $n = m$. $\mathbb{F}_n$ *is minimal except if* $n = 1$; $\mathbb{F}_1$ *is isomorphic to* $\mathbb{P}^2$ *with a point blown up.*

Proof (i) follows from III.18. To prove (ii), consider the section s of the projection $\mathbb{F}_n \to \mathbb{P}^1$ which corresponds to the quotient $\mathcal{O}_{\mathbb{P}^1}$ of $\mathcal{O}_{\mathbb{P}^1} \oplus \mathcal{O}_{\mathbb{P}^1}(n)$ (see III.17). Set $B = s(C)$ and let b denote its class in $\operatorname{Pic} \mathbb{F}_n$; then $b = h + rf$ for some $r \in \mathbb{Z}$. Since $s^*\mathcal{O}_{\mathbb{F}_n}(1) = \mathcal{O}_{\mathbb{P}^1}$, it follows that $h.b = 0$, so that $r = -n$; then $b^2 = (h - nf)^2 = -n$.

Now let C be an irreducible curve on $\mathbb{F}_n$ with $C \neq B$. Set $[C] = \alpha h + \beta f$ in $\operatorname{Pic} \mathbb{F}_n$. Since $C.f \geqslant 0$, we get $\alpha \geqslant 0$; since $C.B \geqslant 0$ and $h.b = 0$, we get $\beta \geqslant 0$. Then $C^2 = \alpha^2 n + 2\alpha\beta \geqslant 0$, and (ii) follows.

Note that on $\mathbb{F}_0 = \mathbb{P}^1 \times \mathbb{P}^1$, $C^2 \geqslant 0$ for all curves C; it follows that the number n is uniquely defined by $\mathbb{F}_n$, and that $\mathbb{F}_n$ is minimal for $n \neq 1$. Finally, let $0 \in \mathbb{P}^2$, S be the blow-up of $\mathbb{P}^2$ at 0 and E be the exceptional divisor. Projecting away from 0 defines a morphism $p : S \to \mathbb{P}^1$ (Example II.14(1)) which expresses S as a surface geometrically ruled over $\mathbb{P}^1$. Since $E^2 = -1$, $S \cong \mathbb{F}_1$, and (iii) follows.

For a more classical description of the surfaces $\mathbb{F}_n$, see Exercises 1, 2.

IV.2 Examples of Rational Surfaces

Let $S \subset \mathbb{P}^n$ be a rational surface. By choosing a birational map $\mathbb{P}^2 \dashrightarrow S$ we get a rational map $\phi : \mathbb{P}^2 \dashrightarrow \mathbb{P}^n$, and so (II.6) a linear system on $\mathbb{P}^2$ with no fixed component. We shall consider the simplest linear systems on $\mathbb{P}^2$ (conics, cubics) and study the embedded rational surfaces that correspond to them.

Let P be such a system and $\phi : \mathbb{P}^2 \dashrightarrow \check{P} \cong \mathbb{P}^N$ the corresponding rational map. We shall be concerned with the following questions:

(1) Determining the dimension N of P; recall at this point the following fact:

Fact IV.3 $\dim H^0(\mathbb{P}^r, \mathcal{O}_{\mathbb{P}^r}(k)) = \binom{r+k}{r}$.

(2) The map ϕ is in general not everywhere defined; to make it so, we must blow up the base points of the system P. For simplicity, we shall assume that it is enough to blow up once; in other words, if $\epsilon : S \to \mathbb{P}^2$ is the blow-up of r distinct base points $p_1, \ldots, p_r$, then $f = \phi \circ \epsilon : S \to \mathbb{P}^N$ is a morphism. It corresponds to a linear system $\hat{P}$ on S; denoting by m_i the minimum multiplicity of the members of P at P_i, d their degree and setting $L = \epsilon^*\ell$ (ℓ a line in $\mathbb{P}^2$), $E_i = \epsilon^{-1}(p_i)$, we get $\hat{P} \subseteq |dL - \sum m_i E_i|$.

(3) We shall be especially interested in the cases where f is an embedding. From the definition of ϕ, this means that

   (a) the linear system $\hat{P}$ on S separates points: for all x, $y \in S$ with $x \neq y$, there is a curve C in $\hat{P}$ passing through x and not through y;

   (b) $\hat{P}$ separates tangent vectors: for all $x \in S$, the curves in $\hat{P}$ through x do not all have the same tangent directions.

   From the point of view of the system P on $\mathbb{P}^2$, (a) can be interpreted by translating 'C passes through x' (for $x \in E_i$) into 'C passes through p_i with the tangent direction corresponding to x'. If $x \in E_i$, then (b) can be interpreted as follows: let P_x be the system of curves in P tangent along x at p_i. For every conic Q tangent at p_i along x, there is a curve in P_x having contact with Q of order exactly 2 at p_i.

(4) Suppose that f is an embedding; the surface $S' = f(S)$ is then a smooth rational surface in $\mathbb{P}^N$. We shall be concerned with the geometry of S', and more precisely:

(a) its Picard group: with the same notation as in (2), it has an orthogonal basis consisting of $L = \epsilon^*\ell$ and the E_i, with $L^2 = 1$, $E_i^2 = -1$. A hyperplane section H of S' can be written as $dL - \sum m_i E_i$, where d is the degree of a curve in P;

(b) its degree; this equals $H^2 = d^2 - \sum_i m_i^2$;

(c) the lines on S'; these are the curves D with $D.H = 1$;

(d) possibly the equations defining S' in $\mathbb{P}^N$.

(5) Let H be a hyperplane in the linear system P, corresponding to a point $h \in \check{P}$. The linear system H defines a rational map $\psi : S \dashrightarrow \check{H}$; one checks readily that this is the composite of ϕ and the projection of $\check{P}$ onto $\check{H}$ away from h. In particular, if $s \in S'$, then the rational surface obtained from the system of curves in P through s is the projection of S' away from $f(s)$.

We are thus led to consider the projections of S' into spaces of dimension $< N$. At this point recall the following lemma, whose proof is immediate.

Lemma IV.4 *Let S be a surface in $\mathbb{P}^N$ and $p \in \mathbb{P}^N - S$ (resp. $p \in S$). Let $f : S \to \mathbb{P}^{N-1}$ (resp. $\hat{S} \to \mathbb{P}^{N-1}$) be the projection away from p. For f to be an embedding, it is necessary and sufficient that there should be no line through p meeting S in at least 2 (resp. 3) points, counted with multiplicity.*

The set of bisecants (resp. tangents) to S is parametrized by the complement of the diagonal in $S \times S$ (resp. the projective tangent bundle to S). It follows that the union of the bisecants (resp. tangents) to S lies in a subvariety of $\mathbb{P}^N$ of dimension $\leqslant 5$ (resp. $\leqslant 4$). Then Lemma IV.4 gives the following result:

Proposition IV.5 *Every surface is isomorphic, via generic projection, to a smooth surface in $\mathbb{P}^5$.*

If S is a surface in $\mathbb{P}^5$, then projecting away from a generic point is an embedding outside a finite set of points $p_1, p_1', \ldots, p_s, p_s'$, and identifies p_i and p_i'.

Now return to the linear systems on $\mathbb{P}^2$.

Linear Systems of Conics

It is clear that the linear system of all conics on $\mathbb{P}^2$ defines an embedding $j : \mathbb{P}^2 \hookrightarrow \mathbb{P}^5$. The image of j is the Veronese surface V, of degree 4. It contains no lines (for if $d \in \operatorname{Pic} V$ were the class of a line, we would have $h.d = 2\ell.d = 1$). On the other hand, it contains a 2-dimensional linear system of conics, which are the images of the lines on $\mathbb{P}^2$. This property has an amusing consequence.

Proposition IV.6 *Let p be a generic point of $\mathbb{P}^5$. Then projecting away from p induces an isomorphism of V onto its image $V' \subset \mathbb{P}^4$.*

(In fact one can show that the Veronese surface is the only surface in $\mathbb{P}^5$ with this property: see [Se2].)

Proof For every line d in $\mathbb{P}^2$, let $P(d)$ denote the plane containing the conic $j(d)$, and let X be the union of the $P(d)$. Then $\dim X \leqslant 4$; for it is the projection to $\mathbb{P}^5$ of the variety $Z \subset \check{\mathbb{P}}^2 \times \mathbb{P}^5$ defined by $Z = \{(d, x) \mid x \in P_d\}$, and Z is a $\mathbb{P}^2$-bundle over $\check{\mathbb{P}}^2$. Let $x, y \in V$ with $x \neq y$; the line $\langle x, y \rangle$ lies in $P(d)$, where d is the line $\langle j^{-1}(x), j^{-1}(y) \rangle$ in $\mathbb{P}^2$. Hence every bisecant of V lies in X, and the result follows.

The generic projection of the surface V' is of degree 4 in $\mathbb{P}^3$, and is called a Steiner surface. One can show (Exercise 7) that it has 3 double lines meeting in a point that is triple on the surface.

The projection of V from a point of V is the surface $S \subset \mathbb{P}^4$ of degree 3 obtained from the linear system of conics on $\mathbb{P}^2$ passing through a point 0. It follows that this system defines an embedding $j : \mathbb{F}_1 \hookrightarrow \mathbb{P}^4$, where $\mathbb{F}_1$ denotes the blow-up of $\mathbb{P}^2$ at 0 (IV.1(iii)). With the notation of IV.1, j corresponds to the linear system $|h + f|$ on $\mathbb{F}_1$.

We now look for the lines on S. We have $f.(h + f) = 1$, so that the images of the fibres of the ruling $\mathbb{F}_1 \to \mathbb{P}^1$ form a family of lines $\{D_t\}_{t \in \mathbb{P}^1}$ on S, with $D_t \cap D_{t'} = \emptyset$ for $t \neq t'$. Since $b.(h + f) = 1$, the image of B (IV.1(ii)) is a line which meets every line D_t. These are the only lines on S; every irreducible curve C on $\mathbb{F}_1$ other than B is equivalent (in $\operatorname{Pic} \mathbb{F}_1$) to $ah + bf$ with $a, b \geqslant 0$ (IV.1(ii)). Thus $C.(h + f) = 2a + b > 1$ unless $C = f$.

The same argument shows that the conics in S are the images under j of the total transforms of the lines in $\mathbb{P}^2$; the lines that do not contain 0 give smooth conics, while those through 0 give the degenerate conics $j(B) \cup D_t$.

Proposition IV.7 *The cubic ruled surface $S \subset \mathbb{P}^4$ is contained in a 2-dimensional linear system of quadrics in $\mathbb{P}^4$, of which it is the intersection. For every pencil of quadrics $\{\lambda Q_1 + \mu Q_2\}$ containing S, $Q_1 \cap Q_2 = S \cup P$, where P is a plane and $S \cap P$ a conic. Conversely, for every conic on S lying in a plane P, $S \cup P$ is the intersection of two quadrics.*

Proof The quadrics of $\mathbb{P}^4$ cut out on S the strict transforms of the quartics in $\mathbb{P}^2$ that pass through 0 with multiplicity 2. The linear systems $|\mathcal{O}_{\mathbb{P}^4}(2)|$ and $|\mathcal{O}_{\mathbb{P}^2}(4)|$ have the same dimension; since passing doubly through a given point imposes three conditions on a system of plane curves, there are at least three linearly independent quadrics containing S. Let Q_1, Q_2 be two of them; they must be irreducible. $Q_1 \cap Q_2$ is a surface of degree 4 containing S, and so is of the form $S \cup P$, for some plane P. If P is defined by the equations $L = M = 0$, then the equation of Q_i is of the form $LA_i + MB_i = 0$ $(i = 1, 2)$, where L, M, A_i, B_i are linear forms. The determinant $A_1B_2 - A_2B_1$ vanishes at every point of $S - P$, and so S lies in the quadric $Q_3 : A_1B_2 - A_2B_1 = 0$. One checks easily that $S = Q_1 \cap Q_2 \cap Q_3$, so that the linear system R of quadrics containing S is generated by Q_1, Q_2, Q_3 and so is 2-dimensional. The intersection $P \cap S = P \cap Q_3$ is a conic. Conversely, if C is a conic in S and P the plane containing C, then there is a pencil of quadrics containing $S \cup P$; indeed, a quadric in R contains P if and only if it contains a point of $P - C$. It is then clear that the intersection of the members of this pencil is $S \cup P$.

Corollary IV.8 *The projection of S from a point $p \in \mathbb{P}^4 - S$ is a cubic surface whose only singularities are a double line.*

Proof Let Q_1, Q_2 be distinct quadrics containing S and p. We have $Q_1 \cap Q_2 = S \cup P$, where P is a plane through p. Every bisecant of S through p cuts Q_1 and Q_2 in 3 points; thus it lies in Q_1 and Q_2, and so also in P. Hence the projection away from p is an isomorphism outside the conic $C = S \cap P$, and its restriction to C is two-to-one onto a line, which is thus double on $p(S)$.

The projection of S from a point $s \in S$ is a quadric in $\mathbb{P}^3$; it is smooth if and only if $s \notin j(B)$.

Finally, we can display the surfaces we have obtained by means of the following diagram, where vertical (resp. diagonal) arrows denote projection from a generic point in space (resp. in the surface):

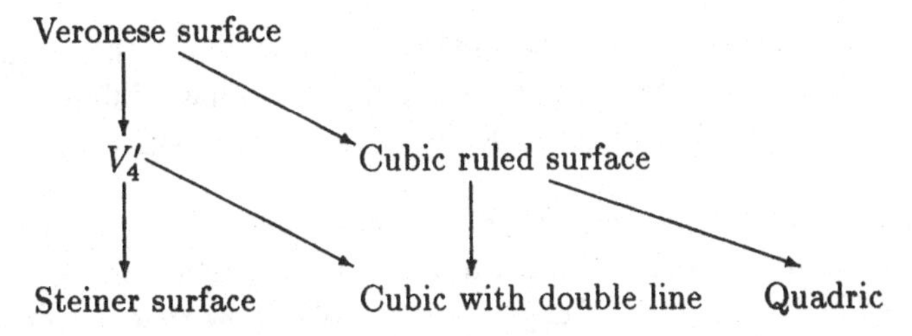

Linear Systems of Cubics

We shall consider r distinct points $p_1, \cdots, p_r \in \mathbb{P}^2$ ($r \leqslant 6$). These points will be said to be in general position if no 3 of them are collinear and no six lie on a conic. $\epsilon : P_r \to \mathbb{P}^2$ will denote the blow-up of $p_1, \ldots, p_r$. Set $d = 9 - r$.

Proposition IV.9 *Suppose that $p_1, \ldots, p_r$ ($r \leqslant 6$) are in general position. Then the linear system of cubics through $p_1, \ldots, p_r$ defines an embedding $j : P_r \hookrightarrow \mathbb{P}^d$. The surface $S_d = j(P_r)$ is a surface of degree d in $\mathbb{P}^d$, called a del Pezzo surface of degree d. In particular, S_3 is a cubic in $\mathbb{P}^3$ and S_4 is the complete intersection of two quadrics in $\mathbb{P}^4$.*

(See [X1] for the case when $p_1, \ldots, p_r$ are not in general position.)

Proof By IV.2(3), we must check that the system of cubics through $p_1, \ldots, p_r$ separates points and tangent vectors on P_r. Then in particular, the system is without base points on P_r, and by induction on r, its dimension is $9 - r$. Moreover, it is enough to check this for $r = 6$, the other cases following at once.

Let $i < j \leqslant 6$ and $x \in P_6$ be such that $\epsilon(x) \notin \{p_i, p_j\}$. The hypothesis of general position implies that there is a unique conic Q_{ij}^x through x and the points p_k ($k \neq i, j$) (recall that if $x \in E_i$, passing through x means being tangent at p_i in the direction corresponding to x). Similarly there is a unique conic Q_i through the points p_j for $j \neq i$; $\hat{Q}_i \cap \hat{Q}_j = \emptyset$ for $i \neq j$ (the hat denoting strict transform). Let $L_{ij} = \langle p_i, p_j \rangle$ for $i \neq j$.

(a) We want to show that this linear system separates points on P_6. So let x, $y \in P_6$ with $x \neq y$. Choose i with $p_i \neq \epsilon(x)$, $\epsilon(y)$ and $x \notin \hat{Q}_i$. Then $\hat{Q}_{ij}^x \cap \hat{Q}_{ik}^x = \{x\}$ for $p_k \notin \{p_i, p_j, \epsilon(x)\}$. Hence $y \in \hat{Q}_{ij}^x$ for at most one value of j. On the other hand, $y \in \hat{L}_{ij}$ for at most one j; thus there is a j such that the cubic $\hat{Q}_{ij}^x \cup \hat{L}_{ij}$ passes through x but not y. Hence the morphism $j : P_6 \to \mathbb{P}^3$ is injective.

(b) Let $x \in \mathbb{P}^2 - \{p_1, \ldots, p_6\}$. The cubics $Q_i \cup \langle p_i, x \rangle$ do not all have the same tangent at x, so that j is an immersion at x. Now let

$x \in E_1$; the conics Q_{23}^x and Q_{24}^x intersect at x with multiplicity 2. Then the cubics $\hat{Q}_{23}^x \cup \hat{L}_{23}$ and $\hat{Q}_{24}^x \cup \hat{L}_{24}$ have different tangents at x, which completes the proof that j is an embedding.

Following IV.2(4), $\deg(j(P_r)) = 9 - r$, so that $j(P_r) = S_d$ is a smooth surface of degree d in $\mathbb{P}^d$, with $d = 9 - r$. In particular, S_3 is a cubic surface in $\mathbb{P}^3$ and S_4 is of degree 4 in $\mathbb{P}^4$. We show that S_4 is contained in two distinct quadrics. Since $h^0(\mathbb{P}^4, \mathcal{O}_{\mathbb{P}^4}(2)) = 15$, we must show that $h^0(\mathcal{O}_{S_4}(2)) \leqslant 13$. Let $C \in |H|$ be a smooth hyperplane section of S_4; C is an elliptic curve, since it projects isomorphically onto a smooth cubic in $\mathbb{P}^2$. Consider the exact sequence

$$0 \to \mathcal{O}_{S_4}(H) \to \mathcal{O}_{S_4}(2H) \to \mathcal{O}_C(2H) \to 0 ;$$

the corresponding long exact cohomology sequence gives

$$h^0(\mathcal{O}_{S_4}(2)) \leqslant h^0(\mathcal{O}_{S_4}(1)) + h^0(\mathcal{O}_C(2)) .$$

Since $H.C = H^2 = 4$, we have $h^0(\mathcal{O}_C(2)) = 8$ by Riemann–Roch, and so $h^0(\mathcal{O}_{S_4}(2)) \leqslant 13$.

Thus S_4 lies in two distinct quadrics Q_1 and Q_2, which must be irreducible; $Q_1 \cap Q_2$ is then a surface of degree 4 containing S_4, and so equal to it.

Remarks IV.10

(1) Note that the linear system of cubics through $p_1, \dots, p_r$ is in fact the complete 'anticanonical system' $|-K|$ on P_r. One can show that together with $\mathbb{P}^1 \times \mathbb{P}^1$ embedded in $\mathbb{P}^8$, the del Pezzo surfaces are the only ones embedded in $\mathbb{P}^N$ by their complete anticanonical system (Chapter V, Exercise 2).

(2) Cubics and complete intersections of two quadrics are the only complete intersection surfaces embedded by their anticanonical system. This follows from the next lemma (a form of the adjunction formula).

Lemma IV.11 *Let $S \subset \mathbb{P}^{r+2}$ be a surface that is the complete intersection of hypersurfaces $H_1, \dots, H_r$ of degrees $d_1, \dots, d_r$ respectively. Then $\mathcal{O}_S(K_S) \cong \mathcal{O}_S(\sum d_i - r - 3)$.*

Proof Let I be the ideal defining $S \subset \mathbb{P}^{r+2}$. Since the equations of the H_i generate I, there is a surjection $\mathcal{O}_{\mathbb{P}^{r+2}}(-d_1) \oplus \cdots \oplus \mathcal{O}_{\mathbb{P}^{r+2}}(-d_r) \to I$, and hence a surjection (by restricting to S)

$$u : \mathcal{O}_S(-d_1) \oplus \cdots \oplus \mathcal{O}_S(-d_r) \to I/I^2 .$$

Both sides are locally free of rank r, and so u is an isomorphism. Hence $\wedge^r(I/I^2) \cong \mathcal{O}_S(-\sum d_i)$; since $\Omega^{r+2}_{\mathbb{P}^{r+2}} \cong \mathcal{O}(-r-3)$, the result follows by I.17.

We shall see later that every cubic surface in $\mathbb{P}^3$ and every intersection of two quadrics in $\mathbb{P}^4$ is a del Pezzo surface.

Proposition IV.12 *S_d contains a finite number of lines. They are the images under j of the following curves in P_r:*

(i) *the exceptional curves E_i;*
(ii) *the strict transforms of the lines $\langle p_i, p_j \rangle$ $(i \neq j)$;*
(iii) *the strict transforms of the conics through 5 of the p_i.*

Their numbers are given in this table:

r = no. of E_i	0	1	2	3	4	5	6
No. of lines $\langle p_i, p_j \rangle$	0	0	1	3	6	10	15
No. of conics through 5 of the p_i	0	0	0	0	0	1	6
No. of lines in S_d	0	1	3	6	10	16	27

Proof Since $H \equiv -K$, the lines on S are just its exceptional curves; in particular, the $j(E_i)$ are lines on S. Let E be a line on S other than an E_i; $E.H = 1$ and $E.E_i = 0$ or 1. Hence $E \equiv mL - \sum m_i E_i$, with $m_i = 0$ or 1 and $E.H = 3m - \sum m_i = 1$. The only solutions are $m = 1$ with 2 of the m_i equal to 1, and $m = 2$ with 5 of the m_i equal to 1; this gives the curves in the statement. Checking the numbers of the curves is immediate.

This proposition in fact gives much more than just the number of lines on S; it gives their classes in $\operatorname{Pic} S_d$, and so the configuration of the set of lines (incidence, etc.). These configurations were studied intensively by the classical geometers – cf. Exercises 12, 13, 14.

Theorem IV.13 *Let $S \subset \mathbb{P}^3$ be a smooth cubic surface. Then S is a del Pezzo surface S_3 (i.e. is isomorphic to $\mathbb{P}^2$ with 6 points blown up).*

First we prove two lemmas.

Lemma IV.14 *S contains a line.*

Proof Let $P = |\mathcal{O}_{\mathbb{P}^3}(3)|$ be the projective space of cubics of $\mathbb{P}^3$, and let G_4 denote the Grassmannian of lines in $\mathbb{P}^3$ (since a line in $\mathbb{P}^3$ is given by 6 Plücker coordinates, subject to a single quadratic relation, G_4 is naturally a smooth quadric in $\mathbb{P}^5$, and is 4-dimensional). Let Z denote the incidence subvariety $Z = \{(\ell, S) \mid \ell \subset S\} \subset G_4 \times P$, with its projections $p : Z \to G_4$ and $q : Z \to P$. In coordinates (X, Y, Z, T) for $\mathbb{P}^3$, a cubic surface $S \subset \mathbb{P}^3$ contains the line $(Z = T = 0)$ if and only if the 4 coefficients of X^3, X^2Y, XY^2, Y^3 in its defining equation F vanish; hence the fibres of $p : Z \to G_4$ have dimension $\dim P - 4$, and $\dim Z = \dim P$. If the lemma fails, the image $q(Z) \subset P$ has codimension $\geqslant 1$ in P, and the fibre $q^{-1}(S)$ is either empty or positive-dimensional for any $S \in P$; thus q is surjective provided that we can display a cubic S containing a (non-empty) finite set of lines. But this is the case if S is a del Pezzo surface S_3.

Lemma IV.15 *Let $\ell \subset S$ be a line; then there are exactly 10 other lines in S meeting ℓ (and distinct from ℓ). These fall into 5 disjoint pairs of concurrent lines. In particular, S contains two disjoint lines.*

Proof Consider the pencil of planes $\{P_\lambda\}_{\lambda \in \mathbb{P}^1}$ through ℓ. One has $S \cap P_\lambda = \ell \cup C_\lambda$, with C_λ a conic. Note that C_λ cannot be a double line, and does not contain ℓ: for if the plane $L = 0$ cuts S along the line $M = 0$ and the line $N = 0$ counted twice, then the equation of S can be written $LQ + MN^2 = 0$ (with Q a quadratic form, L, M, N linear forms). But then, by computing the derivatives, S is singular at the two points where $L = N = Q = 0$. Thus each singular conic C_λ is a union of two concurrent lines, distinct from ℓ but meeting it, and the singular C_λ give all lines of S meeting ℓ.

Choose coordinates on $\mathbb{P}^3$ such that ℓ is given by $Z = T = 0$. Then S is given by an equation of the form

$$AX^2 + 2BXY + CY^2 + 2DX + 2EY + F = 0,$$

where $A, \ldots, F$ are homogeneous polynomials in Z, T. Setting $Z = \lambda T$ and dividing by T gives the equation of C_λ. Thus C_λ is singular just when the determinant

$$\Delta(Z, T) = \begin{vmatrix} A & B & D \\ B & C & E \\ D & E & F \end{vmatrix}$$

vanishes. This is a homogeneous quintic in Z, T and we must show that it has no repeated root. Suppose that $Z = 0$ is a root; let s be the singular point of the conic C_0. If $s \notin \ell$, then we can assume that C_0 is defined by $XY = 0$; then every coefficient of Δ except B is divisible by Z. s is a smooth point of S, and so F is not divisible by Z^2. Hence Δ is not divisible by Z^2. If $s \in \ell$, then we can assume that C_0 is defined by $X^2 - T^2 = 0$; the same argument shows that $Z = 0$ is a simple root of Δ. Therefore $\{C_\lambda\}$ contains 5 distinct singular conics.

To prove the final statement in the lemma, note that if 3 distinct lines in S meet at a point p, then they lie in the tangent plane to S at p, and so are coplanar. Then let C_0, C_1 be singular conics in the pencil $\{C_\lambda\}$, $C_0 = d_0 \cup d'_0$, $C_1 = d_1 \cup d'_1$; since d_0, d_1 and ℓ are not coplanar, it is clear that $d_0 \cap d_1 = \emptyset$.

Proof of Theorem IV.13 Let ℓ, ℓ' be disjoint lines in S. We define rational maps $\phi : \ell \times \ell' \dashrightarrow S$ and $\psi : S \dashrightarrow \ell \times \ell'$ as follows: if (p, p') is a generic point of $\ell \times \ell'$, the line $\langle p, p' \rangle$ meets S in a third point p''; define $\phi(p, p') = p''$. For $s \in S - \ell - \ell'$, set $p = \ell \cap \langle s, \ell' \rangle$, $p' = \ell' \cap \langle s, \ell \rangle$ and put $\psi(s) = (p, p')$. It is clear that ϕ, ψ are mutually inverse. Moreover, ψ is a morphism; it can be defined on ℓ (or ℓ') by replacing the plane $\langle s, \ell \rangle$ by the tangent plane to S at s (checking immediately that this gives a morphism). Thus ψ is a birational morphism, and so a composite of blow-ups; the curves contracted by ψ are just those lines that meet ℓ and ℓ'.

We proceed to calculate the number of these lines. We know (Lemma IV.15) that the lines meeting ℓ fall into 5 pairs $\{d_i, d'_i\}$ $(1 \leqslant i \leqslant 5)$ such that d_i, d'_i and ℓ lie in a plane P_i. P_i meets ℓ' in one point, which lies on d_i or d'_i (but not both, for else d_i, d'_i and ℓ' would be coplanar). Thus one line in each pair meets ℓ' as well, and so ψ contracts 5 disjoint lines. Hence S is isomorphic to $\mathbb{P}^1 \times \mathbb{P}^1$ with five points (p_i, p'_i) blown up, with $p_i \neq p_j$ and $p'_i \neq p'_j$ for $i \neq j$; in view of II.14(2), S is then isomorphic to $\mathbb{P}^2$ with six points blown up. Moreover the embedding of S in $\mathbb{P}^3$ is the anticanonical embedding (Remark IV.10(2)), and so S is the image of the blow-up of $\mathbb{P}^2$ at six points embedded by the system of cubics through the six points; in other words, S is a del Pezzo surface S_3.

Proposition IV.16 *Let S be the complete intersection of two quadrics in $\mathbb{P}^4$. Then S is a del Pezzo surface S_4.*

Proof We show first that S contains only finitely many lines. Any line $E \subset S$ satisfies $E.H = 1$, so $E.K = -1$ and $E^2 = -1$; hence distinct lines E, E' have different classes in $NS(S)$ (since $E.E' = 0$ or 1). Fix a line $\ell \subset S$; if two distinct lines E, E' met ℓ, then $E.E' = 0$ (or else the plane containing E, E', ℓ would lie in S). Since $NS(S)$ is finitely generated, there are only finitely many lines in S meeting a given one. If S contained infinitely many lines, then we could construct an infinite sequence $\{E_n\}$ of lines with $E_i.E_j = 0$ for $i \neq j$, in contradiction to the finiteness of $NS(S)$.

Suppose that $p \in S$ does not lie on any line of S and let $f : \hat{S} \to \mathbb{P}^3$ be the projection away from p. Every trisecant to S lies in the quadrics containing S, and so in S. In particular, there is no trisecant through p. By IV.4, f is an isomorphism onto a cubic surface $S_3 \subset \mathbb{P}^3$. The exceptional divisor of the blow-up at p gives a line $E \subset S_3$. Choosing two disjoint lines ℓ, ℓ' that meet E, as above, we get a birational morphism $h : S_3 \to \mathbb{P}^2$ contracting six lines, including E. Thus h factors through a morphism $S \to \mathbb{P}^2$ so that S is isomorphic to $\mathbb{P}^2$ with five points blown up. Since S is embedded in $\mathbb{P}^4$ by its complete anticanonical system, it is a del Pezzo surface S_4.

Historical Note IV.17

The examples in this chapter (and the exercises) are only a tiny sample of the huge mass of results – now practically forgotten – on particular surfaces proved by nineteenth century geometers (Clebsch, Cremona, Darboux, Klein, Kronecker, Kummer, ...). The point of view taken before 1880 was slightly different from ours, since they only considered (possibly singular) surfaces in $\mathbb{P}^3$. The 'geometry of hyperspace' arose with the first generation of the Italian school (Bertini, C. Segre, Veronese); they discovered that many known examples of surfaces are projections of smooth surfaces embedded in a higher dimensional space, and that this gives a natural explanation of many of their properties.

For example, Veronese ([V]) introduced the surface named after him, studied its projections and so rediscovered Steiner's 'Roman surface' (found in 1844). C. Segre showed that a quartic in $\mathbb{P}^3$ with a double conic is the projection of a quartic del Pezzo surface in $\mathbb{P}^4$ (Exercise 15), and thus easily derived its properties ([Sg2]). He also made a systematic study of ruled surfaces ([Sg1]).

Del Pezzo classified surfaces of degree n in $\mathbb{P}^n$ and at the same time introduced the surfaces S_d ([DP]).

For an excellent exposition of this subject, see *Le superficie razionali*, by Conforto (following Enriques), Zanichelli, Bologna, 1939.

Exercises IV.18

(1) With the notation of IV.1, show that the complete linear system $|h|$ on $\mathbb{F}_n$ defines a morphism $f : \mathbb{F}_n \to \mathbb{P}^{n+1}$ which is an embedding outside B and contracts B to a point p. Show that $f(\mathbb{F}_n)$ is the cone with vertex p over a projectively normal rational curve of degree n in $\mathbb{P}^n$ (i.e. $\mathbb{P}^1$ embedded by the complete system $|\mathcal{O}(n)|$).
(Show that $h^1(h-f)=0$, for example by comparing it to $h^1(-f)$; also use the exact sequence

$$0 \to \mathcal{O}_{\mathbb{F}_n} \to \mathcal{O}_{\mathbb{F}_n}(C) \to \mathcal{O}_C(C) \to 0, \qquad C \in |h|\,.)$$

(2) Show that $|h+kf|$ $(k \geqslant 1)$ defines an embedding $j : \mathbb{F}_n \hookrightarrow \mathbb{P}^{n+2k+1}$. Show that the fibres f_t are mapped to a family of disjoint lines; the curve $j(B)$ (resp. $j(C)$, for generic $C \in |h|$) is a projectively normal rational curve of degree k (resp. $n+k$) which meets each line $j(f_t)$ once; $j(\mathbb{F}_n)$ is of degree d in $\mathbb{P}^{d+1}$, with $d = n+2k$.
Conversely, let H_k and H_{d-k} be disjoint linear spaces in $\mathbb{P}^{d+1}$, of dimension k and $d-k$ respectively, with $2k \leqslant d$; let R_k (resp. R_{d-k}) be a projectively normal rational curve of degree k (resp. $d-k$) in H_k (resp. H_{d-k}), and let $u : R_k \xrightarrow{\sim} R_{d-k}$ be an isomorphism. Show that $\bigcup_{r \in R_k} \langle r, u(r) \rangle$ is a copy of $\mathbb{F}_n$ $(n = d-2k)$, embedded by the system $|h+kf|$.

(3) Show that every irreducible surface (possibly singular) of degree $\leqslant n-2$ in $\mathbb{P}^n$ lies in a hyperplane.

(4) Let $S \subset \mathbb{P}^n$ be a surface (possibly singular) of degree $n-1$, not lying in a hyperplane. Show that S is one of the following:

  (a) a cone over a projectively normal rational curve of degree $n-1$ in $\mathbb{P}^{n-1}$;

  (b) the Veronese surface;

  (c) the surface $\mathbb{F}_r$ embedded by $|h+kf|$, where $r = n-1-2k$, $k \geqslant 1$.

(If S is singular, it is a cone by (3).) If it is smooth, show that a smooth hyperplane section H of S is rational, then that the linear system $|K_S + 2H|$ is base-point free; deduce that either $K_S^2 = 9$ and we are in case (b), or $K_S^2 = 8$ and we are in case (c).

(4) Let $P = \mathbb{P}^2$ and $\check{P}$ the dual plane. The projective space Q of conics in P is dual (by 'apolarity') to the space $\check{Q}$ of conics in $\check{P}$. Let $q \in \check{Q}$, and show that:

  (a) If q is of rank 1 (i.e. q is the set of lines through a point p of P), then the conics in P apolar to q are those through p.

(b) If q is of rank 2 (i.e. q is the set of lines passing through one of two points $p_1, p_2 \in P$), then a conic C in P is apolar to q if and only if the two points of $C \cap \langle p_1, p_2 \rangle$ are harmonic conjugates with respect to p_1, p_2.

(5) The system of conics defines an embedding $P \hookrightarrow \check{Q}$ whose image is the Veronese surface $V \subset \check{Q}$. Show that V is the set of conics of rank 1 in $\check{P}$ (use (5)(a)). Deduce that V is the intersection of 5 quadrics in $\check{Q}$. Show that the union of the bisecants of V is the set X of singular conics in $\check{P}$. Deduce that X is a cubic hypersurface in $\check{Q}$ whose singular locus is V.

(6) Let $f : \mathbb{P}^2 \to \mathbb{P}^3$ be the morphism corresponding to a 3-dimensional base-point free linear system of conics. Let $\check{R}$ be the dual pencil of conics in $\check{\mathbb{P}}^2$; suppose that $\check{R}$ contains 3 distinct singular conics, each of rank 2. These correspond to 3 pairs of points (p_i, p_i'), $i = 1, 2, 3$. Show that f is an embedding outside the 3 lines $\langle p_i, p_i' \rangle$, maps each of these lines two-to-one onto a double line in $S = f(\mathbb{P}^2)$ and takes the vertices of the triangle formed by these lines to a single triple point of S, the point of intersection of the 3 double lines of S. S is Steiner's 'Roman surface'.

(7) Let $S \subset \mathbb{P}^3$ be a quartic surface with 3 non-coplanar double lines meeting in a triple point t of S. Show that S is a Steiner surface (project away from t and then perform a quadratic transformation).

(8) Let $S \subset \mathbb{P}^3$ be a cubic surface containing a double line d. Show that S is the projection of a cubic ruled surface $R \subset \mathbb{P}^4$. Assume that the conic $C \subset R$ that maps onto d is smooth. Show that there is a line $\ell \subset S$, disjoint from d, and a two-to-one map $f : \ell \to d$ such that $S = \bigcup_{p \in \ell} \langle p, f(p) \rangle$.

(9) Find the number of pencils of conics (resp. systems of twisted cubics) on a del Pezzo surface S_d.

(10) Consider r points in $\mathbb{P}^2$ in general position (if $r = 8$, we require also that they do not all lie on a nodal or cuspidal cubic with one of them at the singular point). Let P_r denote the blow-up of $\mathbb{P}^2$ at these points and H the strict transform of a cubic containing them.

(a) If $r = 7$, show that $|H|$ defines a two-to-one map $P_7 \to \mathbb{P}^2$ branched along a smooth quartic, and that $|2H|$ defines an embedding $P_7 \hookrightarrow \mathbb{P}^6$.

(b) If $r = 8$, show that $|2H|$ defines a two-to-one map from P_8 onto a quadric cone in $\mathbb{P}^3$, and that $|3H|$ defines an embedding $P_8 \hookrightarrow \mathbb{P}^6$.

(11) 'Double-six'. Show that on a cubic surface $S \subset \mathbb{P}^3$ there exists a double-six, that is a set of 12 lines $\ell_1, \dots, \ell_6, \ell'_1, \dots, \ell'_6$ such that $\ell_i \cap \ell_j = \ell'_i \cap \ell'_j = \emptyset$ and $\ell_i \cap \ell'_j = \{pt\}$ for $i \neq j$, $\ell_i \cap \ell'_i = \emptyset$. Show that a cubic surface has 36 double-sixes.

(12) Show that the ten lines on a quintic del Pezzo surface S_5 are arranged as follows: six sides of a skew hexagon; three transversals joining opposite sides of the hexagon; one line joining the three transversals.

(13) It follows from Exercise 11 or a direct computation that the projection of a cubic surface S_3 away from a point $p \in S_3$ is the double cover of $\mathbb{P}^2$ ramified along a smooth quartic C. Show that the lines on S_3 and the exceptional divisor of the blow-up at p map onto bitangents to C; deduce that C has just 28 bitangents.
(Conversely, if we knew that C has exactly 28 bitangents, which is a classical result, we could deduce directly that S_3 contains 27 lines.)

(14) Show that the projection of a quartic del Pezzo surface $S_4 \subset \mathbb{P}^4$ away from a generic point of $\mathbb{P}^4 - S_4$ is a quartic containing a double conic. Conversely, every such surface is the projection of some S_4. Deduce that a quartic in $\mathbb{P}^4$ with a double conic contains 16 lines.

In the next exercises we take distinct points $\{p_i\}$ in $\mathbb{P}^2$ and a linear system P of curves through the $\{p_i\}$. As in IV.2, we want to study the rational map defined by P, under any necessary hypotheses of 'general position'.

(15) $P = \{$curves of degree n through $p_1, \dots, p_{n-1}$, that pass through p_0 with multiplicity $n-1\}$. Show that the image is a surface $\mathbb{F}_r$ embedded by $|h + kf|$, where $r + 2k = n$. The number r depends upon the position of the p_i; if for example $p_1, \dots, p_{n-1}$ are collinear, then $r = n - 2$.

(16) $P = \{$quartics through $p_1, \dots, p_7$, with a double point at $p_0\}$. The image is a quintic surface $S \subset \mathbb{P}^4$; there is a plane H such that $S \cup H$ is the intersection of a quadric and a cubic. C contains 14 lines and a single pencil of conics.

(17) $P = \{$quartics through $p_1, \dots, p_8$, with a double point at $p_0\}$. The image is a quartic surface $S \subset \mathbb{P}^3$ with a double line, the image of a cubic in $\mathbb{P}^2$ through the p_i.

(18) P = {quartics through 9 points}. The image is a degree 7 surface $S \subset \mathbb{P}^5$; there is a plane H such that $S \cup P$ is the complete intersection of 3 quadrics.

(19) P = {quartics through 10 points}. The image is a sextic surface $S \subset \mathbb{P}^4$ (a 'Bordiga surface'). S contains 10 disjoint lines and 10 disjoint plane cubics such that each line meets a single cubic (a 'double-ten').

V

CASTELNUOVO'S THEOREM AND ITS APPLICATIONS

Theorem V.1 (Castelnuovo's Rationality Criterion) *Let S be a surface with $q = P_2 = 0$. Then S is rational.*

Remark V.2 The condition $P_2 = 0$ implies $p_g = 0$. Later we will see that the condition given in the statement cannot be replaced by the weaker condition $q = p_g = 0$, which seems more natural (cf. Enriques surfaces, Godeaux surfaces, ...).

Castelnuovo's theorem has an important corollary. In order to state it we need two definitions:

Definition V.3 *Let V be a variety of dimension n.*

- *V is unirational if there exists a dominant (i.e. generically surjective) rational map $\mathbb{P}^n \dashrightarrow V$.*
- *V is rational if there exists a birational map $\mathbb{P}^n \overset{\sim}{\dashrightarrow} V$. In other words, V is rational (resp. unirational) if the field of rational functions of V is (resp. is contained in) a pure transcendental extension of $\mathbb{C}$.*

Recall that for curves we have:

Theorem V.4 (Lüroth) *Every unirational curve is rational.*

Proof If C is unirational, there is a surjective morphism $f : \mathbb{P}^1 \to C$. There is no non-zero holomorphic form on C (for its inverse image would be a non-zero holomorphic form on $\mathbb{P}^1$); so C is of genus 0, hence rational.

Lüroth's theorem is in fact true over a field (not necessarily algebraically closed) of any characteristic. This is not the case for the analogous result for surfaces:

Corollary V.5 (of Castelnuovo's theorem) *Every unirational surface is rational.*

Proof Let S be a unirational surface. By the theorem on elimination of indeterminacy (II.7), there exists a surjective morphism $R \to S$ where R is a rational surface. Since $q = P_2 = 0$ for R (III.21), we conclude as before that $q = P_2 = 0$ for S, whence the result.

The Lüroth problem for varieties of dimension > 2 remained open for a long time, or rather was not satisfactorily settled: a number of counterexamples were proposed (Fano, Roth, ...), but the proofs of irrationality given are nowdays considered incomplete. Recently some castiron counterexamples have been given ([C-G] and [I-M]): hypersurfaces of degree 3 (resp. 4) in $\mathbb{P}^4$. Notice that this does not involve pathological constructions, but the simplest possible varieties; morally, 'almost all' unirational varieties of dimension $\geqslant 3$ are irrational.

It is unknown whether there exist numerical conditions, analogous to those of V.1, characterizing unirational varieties.

We will deduce the theorem from the following proposition:

Proposition V.6 *Let S be a minimal surface with $q = P_2 = 0$. Then there exists a smooth rational curve C on S such that $C^2 \geqslant 0$.*

We remark that the proposition is not at all obvious for a rational surface S.

V.7 We first show how the proposition implies Castelnuovo's theorem. From the exact sequence

$$0 \to \mathcal{O}_S \to \mathcal{O}_S(C) \to \mathcal{O}_C(C) \to 0$$

and the vanishing of $H^1(S, \mathcal{O}_S)$, we deduce $h^0(C) = 2 + C^2 \geqslant 2$. Let D be a divisor in $|C|$ other than C. The pencil generated by C and D has no fixed components; after blowing up base points it determines a morphism $\hat{S} \to \mathbb{P}^1$ with one fibre isomorphic to C. By the Noether–Enriques theorem (III.4) it follows that S is rational.

To prove the proposition, we will use the following lemma:

Lemma V.8 *Let S be a minimal surface with $K^2 < 0$. For all $a > 0$, there exists an effective divisor D on S such that $K.D \leqslant -a$, $|K + D| = \emptyset$.*

Proof It suffices to find an effective divisor E on S such that $K.E < 0$. Indeed, there is then a component C of E such that $K.C < 0$. The genus formula (I.15) gives $C^2 \geqslant -1$, and $C^2 = -1$ only if C is an exceptional curve, which is excluded. Thus $C^2 \geqslant 0$; since the product $(aC + nK).C$ eventually becomes negative as n grows, it follows from the useful remark (III.5) that $|aC+nK| = \emptyset$ for n sufficiently large. So there exists n such that $|aC+nK| \neq \emptyset$, $|aC+(n+1)K| = \emptyset$; if $D \in |aC+nK|$, we have $K.D \leqslant -a$ and $|K+D| = \emptyset$.

Let H be a hyperplane section of S. If $K.H < 0$, we can take $E = H$; if $K.H = 0$, the system $|K+nH|$ is non-empty for n sufficiently large, and we can take $E \in |K+nH|$. So we may assume $K.H > 0$. Set $r_0 = K.H/(-K^2)$. Then

$$(H + r_0K)^2 = H^2 + \frac{(K.H)^2}{-K^2} > 0\ , \quad \text{and} \quad (H + r_0K).K = 0\ ,$$

so that if r is rational, greater than r_0 and sufficiently close to r_0, we have

$$(H + rK)^2 > 0\ , \quad (H + rK).K < 0\ , \quad (H + rK).H > 0\ .$$

If $r = p/q$, $(p, q > 0)$, put $D_m = mq(H + rK)$. Then D_m is a divisor satisfying $D_m^2 > 0$, $D_m.K < 0$. By the Riemann–Roch theorem we have $h^0(D_m) + h^0(K - D_m) \to \infty$ as $m \to \infty$.

Since $(K - D_m).H$ is negative for m sufficiently large, we have $|D_m| \neq \emptyset$ for large m; we take $E \in |D_m|$.

V.9 Proof of Proposition V.6

(a) It is enough to show that there exists an effective divisor D on S such that $K.D < 0$, $|K + D| = \emptyset$. For then some component C of D satisfies $K.C < 0$, $|K + C| = \emptyset$; applying Riemann–Roch to $K + C$, we obtain

$$0 = h^0(K + C) \geqslant 1 + \frac{1}{2}(C^2 + C.K) = g(C)\ ,$$

so C is a smooth rational curve. By the genus formula $C^2 \geqslant -1$; if $C^2 = -1$, C is an exceptional curve, which is excluded. Thus $C^2 \geqslant 0$, and the proposition is proved.

We distinguish three cases.

(b) $K^2 < 0$.
The proposition follows from (a) and V.8.

(c) $K^2 = 0$.
Since $P_2 = 0$, we have $h^0(-K) \geqslant 1 + K^2$ by Riemann–Roch, so $|-K| \neq \emptyset$ if $K^2 \geqslant 0$. Let H be a hyperplane section of S. There exists $n \geqslant 0$ such that $|H + nK| \neq \emptyset$, $|H + (n+1)K| = \emptyset$. Let $D \in |H + nK|$; we have $|K + D| = \emptyset$, and $K.D = K.H < 0$ since $|-K| \neq \emptyset$. We conclude via (a).

(d) $K^2 > 0$.
In this case $h^0(-K) \geqslant 2$. Suppose there exists a reducible divisor $D \in |-K|$, $D = A + B$; since $D.K < 0$ we have for example $A.K < 0$, and $|K+A| = |-B| = \emptyset$, so the proposition is proved by (a). So we may assume from now on that every divisor $D \in |-K|$ is irreducible.
Let H be an effective divisor; since $|-K| \neq \emptyset$, there exists $n > 0$ such that $|H + nK| \neq \emptyset$, $|H + (n+1)K| = \emptyset$.
It is necessary to distinguish two subcases.

(d_1) Suppose we can find H, n as above, such that $H + nK \not\equiv 0$. Then let $E \in |H + nK|$, $E = \sum n_i C_i$. We have $K.E = -D.E$ ($D \in |-K|$) and by the useful remark (III.5), $D.E \geqslant 0$ since D is irreducible. Thus $K.C_i \leqslant 0$ for some i. Put $C = C_i$. Then $|K + C| = \emptyset$, whence $g(C) = 0$ (cf. (a)), and $C^2 = -2 - K.C$ (by the genus formula).
If $K.C \leqslant -2$, we get $C^2 \geqslant 0$, and the proposition is proved.
If $K.C = -1$, we get $C^2 = -1$: C must be an exceptional curve, which is excluded.
If $K.C = 0$, then $C^2 = -2$; we calculate $h^0(-K - C)$. Since $h^0(2K + C) \leqslant h^0(K + C) = 0$,

$$\begin{aligned} h^0(-K-C) &\geqslant 1 + \frac{1}{2}[(K+C)^2 + K.(K+C)] \\ &= 1 + \frac{1}{2}(C^2 + 3K.C + 2K^2) \end{aligned}$$

by Riemann–Roch, and hence

$$h^0(-K-C) \geqslant K^2 \geqslant 1 \ .$$

Since $C^2 = -2$, we have $C \not\equiv -K$, so there exists a non-zero effective divisor A such that $A + C \in |-K|$. Thus $|-K|$ contains a reducible divisor, which contradicts our assumption.

(d_2) It remains to consider the case where every effective divisor is a multiple of K, that is $\operatorname{Pic} S = \mathbb{Z}[K]$. From the exact sequence

$$H^1(S, \mathcal{O}_S) \to \operatorname{Pic} S \to H^2(S, \mathbb{Z}) \to H^2(S, \mathcal{O}_S)$$

we deduce that $H^2(S,\mathbb{Z}) \cong \mathbb{Z}[K]$. Thus $b_2 = 1$; by Poincaré duality the intersection form in $H^2(S,\mathbb{Z})$ is unimodular, so $K^2 = 1$. But then Noether's formula, I.14,

$$\chi(\mathcal{O}_S) = \frac{1}{12}(K^2 + \chi_{\text{top}}(S)) = \frac{1}{12}(K^2 + 2 - 2b_1 + b_2)$$

gives $b_1 = -4$, a contradiction: this proves the proposition, and hence also Castelnuovo's theorem.

We remark that only the last part of the proof (d_2) does not extend directly to characteristic p.

Proposition V.6 enables us to find the structure of the minimal models of rational surfaces:

Theorem V.10 *Let S be a minimal rational surface. Then S is isomorphic to $\mathbb{P}^2$ or to one of the surfaces $\mathbb{F}_n$* (IV.1) *for $n \neq 1$.*

Proof Let H be a hyperplane section of S. Consider the set A of smooth rational curves C with $C^2 \geqslant 0$; A is non-empty by Proposition V.6. Let $m = \min\{C^2 \mid C \in A\}$. From the subset A_m of A consisting of curves C with $C^2 = m$ we choose a curve C with $C.H$ minimal for C in A_m.

(a) We show that every divisor $D \in |C|$ is a smooth rational curve. Put $D = \sum n_i C_i$. Note that by the useful remark, $h^0(K+D) = h^0(K+C) = 0$ since $(K+C).C = -2$. Thus $h^0(K+C_i) = 0$ for all i, which proves that every curve C_i is a smooth rational curve (V.9(a)). Since $K.C < 0$, there exists i such that $K.C_i < 0$, which implies $C_i^2 \geqslant 0$, since S is minimal. Put $D' = \sum_{i \neq j} n_j C_j$, so that $D = n_i C_i + D'$ and $D'.C_i \geqslant 0$. We have

$$C^2 = D^2 = n_i^2 C_i^2 + n_i(C_i.D') + D.D' \ .$$

Now $D.D' = C.D' \geqslant 0$, so $m = C^2 \geqslant n_i^2 C_i^2 \geqslant 0$. Then by minimality of m, $C_i^2 = m$. Also $H.C = n_i H.C_i + H.D'$, which gives $n_i = 1$, $H.D' = 0$ (minimality of $H.C$ in A_m), that is $D' = 0$ and $D = C_i$.

(b) We show that $\dim|C| \leqslant 2$. Let p be some point on S, $\mathcal{O}_p$ its local ring, ${}_p$ its maximal ideal. Since $\dim(\mathcal{O}_p/\ {}^2_p) = 3$, the linear system of curves of $|C|$ passing through p with multiplicity $\geqslant 2$ has codimension $\leqslant 3$ in $|C|$; thus it is non-empty if $\dim|C| \geqslant 3$, contradicting (a).

(c) Let $C_0 \in |C|$. Consider the exact sequence:

$$0 \to \mathcal{O}_S \to \mathcal{O}_S(C) \to \mathcal{O}_{C_0}(m) \to 0 \ .$$

Since $H^1(S, \mathcal{O}_S) = 0$, we deduce that $h^0(C) = m + 2$, and that $|C|$ is without base points on C_0; it follows that C has no base points at all. On account of (b), there are now two possibilities:

$m = 0$: then $|C|$ determines a morphism $S \to \mathbb{P}^1$ all of whose fibres are smooth rational curves: S is a geometrically ruled surface over $\mathbb{P}^1$, thus a surface $\mathbb{F}_n$, with $n \neq 1$ since S is minimal.

$m = 1$: $|C|$ determines a morphism $f : S \to \mathbb{P}^2$; for every $p \in \mathbb{P}^2$ the fibre $f^{-1}(p)$ is the intersection of two distinct rational curves in $|C|$; so it is reduced to a point. It follows that f is an isomorphism.

We have now determined the minimal models of rational surfaces and ruled surfaces; to complete the study of minimal models, it remains to prove the uniqueness of the minimal model of a non-ruled surface. We are going to deduce this from Castelnuovo's theorem; first we need to introduce a very useful technical tool, the Albanese variety.

Reminder V.11 Complex tori

A complex torus is a quotient manifold $T = V/\Gamma$, where V is a complex vector space and Γ is a lattice in V (i.e. $\Gamma \otimes_{\mathbb{Z}} \mathbb{R} \xrightarrow{\sim} V$). It is a compact analytic manifold, equipped with the structure of an Abelian group. If T admits an embedding into projective space, we say that it is an Abelian variety.

For every point p of T, the tangent space at p is identified by translation to the tangent space at the origin, which is canonically isomorphic to V. Thus the tangent (resp. cotangent) sheaf of T is canonically isomorphic to the free sheaf $V \otimes_{\mathbb{C}} \mathcal{O}_T$ (resp. $V^* \otimes_{\mathbb{C}} \mathcal{O}_T$). In particular there is an isomorphism $\delta : V^* \xrightarrow{\sim} H^0(T, \Omega^1_T)$, given explicitly as follows: a form $x^* \in V^*$ defines a function on V satisfying $x^*(v+\gamma) = x^*(v) + \text{constant}$, for all $v \in V$, $\gamma \in \Gamma$. This ensures that the differential dx^* defines a form $\delta x^* \in H^0(T, \Omega^1_T)$.

The map $V \to V/\Gamma$ is the universal cover of T; hence $\Gamma = \pi_1(T) = H_1(T, \mathbb{Z})$. The isomorphism $h : \Gamma \to H_1(T, \mathbb{Z})$ is given explicity as follows: to every $\gamma \in \Gamma$ we associate the path $t \mapsto t\gamma$ $(0 \leqslant t \leqslant 1)$. Clearly $\int_{h\gamma} \delta x^* = \int_0^1 d\langle x^*, t\gamma\rangle = \langle x^*, \gamma\rangle$ for $x^* \in V$, $\gamma \in \Gamma$.

Finally we recall the following well known proposition:

Proposition V.12 *Let $T_1 = V_1/\Gamma_1$, $T_2 = V_2/\Gamma_2$ be two complex tori, $u : T_1 \to T_2$ a morphism. Then u is composed of a translation and a* *morphism $a : T_1 \to T_2$ which is a group homomorphism; a is induced by a linear map $\overline{a} : V_1 \to V_2$ such that $\overline{a}(\Gamma_1) \subset \Gamma_2$. In particular a is determined by $u^* : H^0(T_2, \Omega^1_{T_2}) \to H^0(T_1, \Omega^1_{T_1})$.*

Proof The morphism u induces a morphism of universal covers $\overline{u} : V_1 \to V_2$, such that $\overline{u}(x+\gamma) - \overline{u}(x) \in \Gamma_2$ for $x \in V_1$, $\gamma \in \Gamma_1$. So the expression $\overline{u}(x+\gamma) - \overline{u}(x)$ is independent of x. This implies that the partial derivatives of $\overline{u}$ are invariant under translations of Γ_1; so they define functions on T_1 which are holomorphic, hence constant. Thus $\overline{u}$ is an affine map, of the form $x \mapsto \overline{a}(x) + b$, where $\overline{a}$ is a homomorphism from V_1 to V_2, and $b \in V_2$. We must have $\overline{a}(\Gamma_1) \subset \Gamma_2$, which implies that $\overline{a}$ induces a homomorphism $a : T_1 \to T_2$. Clearly a^* is identified, via the identifications $\delta : V_i^* \xrightarrow{\sim} H^0(T_i, \Omega^1_{T_i})$, with the transpose of $\overline{a}$, and this completes the proof of the proposition.

Theorem V.13 *Let X be a smooth projective variety. There exists an Abelian variety A and a morphism $\alpha : X \to A$ with the following universal property:*

> *for any complex torus T and any morphism $f : X \to T$, there exists a unique morphism $\tilde{f} : A \to T$ such that $\tilde{f} \circ \alpha = f$.*

The Abelian variety A, determined up to isomorphism by this condition, is called the Albanese variety of X, and written $\mathrm{Alb}(X)$. *The morphism α induces an isomorphism $\alpha^* : H^0(A, \Omega^1_A) \to H^0(X, \Omega^1_X)$.*

Proof We will assume the following result, which is proved by Hodge theory (cf. [W]):

Let $i : H_1(X, \mathbb{Z}) \to H^0(X, \Omega^1_X)^*$ be the map defined by $\langle i(\gamma), \omega \rangle = \int_\gamma \omega$ for $\gamma \in H_1(X, \mathbb{Z})$, $\omega \in H^0(X, \Omega^1_X)$. The image of i is a lattice in $H^0(X, \Omega^1_X)^*$, and the quotient is an Abelian variety.

We write $H = \mathrm{Im}(i)$, $\Omega = H^0(X, \Omega^1_X)$, and put $A = \Omega^*/H$. Next we define α.

Fix a point p in X. Let c_x be a path joining p to a point x in X, and let $a(c_x) \in \Omega^*$ be the linear form $\omega \mapsto \int_{c_x} \omega$. If we replace c_x by another path c'_x joining p to x, we change $a(c_x)$ by an element of H. So the class of $a(c_x)$ in A depends only on x: we call it $\alpha(x)$.

We will show that α is analytic in a neighbourhood of a point $q \in X$. Choose a path c from p to q, and a neighbourhood U of q in X isomorphic to a ball B in $\mathbb{C}^n$; we identify U with B. For $x \in U$, put $a(x) = a(c_x)$, where c_x is the path composed of the path c and the segment $\langle q, x \rangle$. It is clear that $a : U \to \Omega^*$ is an analytic morphism; since $\alpha_{|U} = \pi \circ a$, where

π denotes the projection of Ω^* onto $A = \Omega^*/H$, α is analytic in U. We have $\alpha(p) = 0$; notice that if we change p, α is altered by translation in A.

We prove the last part of the theorem. Since $\delta : \Omega \to H^0(A, \Omega^1_A)$ is an isomorphism (V.11), it is enough to show that $\alpha^*(\delta\omega) = \omega$ for all $\omega \in \Omega$. Locally on X, we can write $\alpha = \pi \circ a$ as above, whence

$$\alpha^*(\delta\omega) = a^*\pi^*(\delta\omega) = a^*d(\langle\omega, \cdot\rangle) .$$

The value of this form at a point $x \in X$ is

$$d(\langle\omega, a(x)\rangle) = d\left(\int_p^x \omega\right) = \omega(x) ,$$

which proves that $\alpha^*(\delta\omega) = \omega$.

We now demonstrate the universal property of A. Let $T = V/\Gamma$ be a complex torus, $f : X \to T$ a morphism. We prove the uniqueness of $\tilde{f}$. There is a commutative diagram

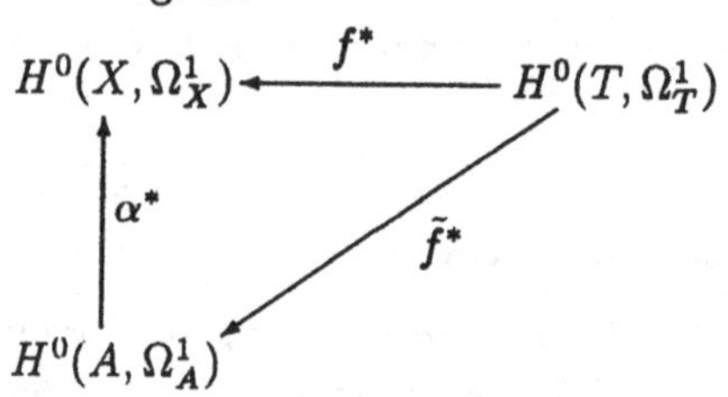

which determines $\tilde{f}^*$, because α^* is an isomorphism. It follows that $\tilde{f}$ is determined up to translation (Proposition V.12); since we fixed $\tilde{f}(0) = f(p)$, $\tilde{f}$ is unique.

To prove the existence of $\tilde{f}$, it suffices (given Proposition V.12) to prove that the composite homomorphism $u : V^* \xrightarrow{\delta} H^0(T, \Omega^1_T) \xrightarrow{f^*} \Omega$ satisfies ${}^t u(H) \subset T$. Let $\gamma \in H_1(X, \mathbb{Z})$, $v^* \in V^*$; then

$$\langle {}^t u(i(\gamma)), v^*\rangle = \langle i(\gamma), u(v^*)\rangle = \int_\gamma f^*(\delta v^*) = \int_{f_*\gamma} \delta v^*$$

and $\int_{f_*\gamma} \delta v^* = \langle h^{-1}(f_*\gamma), v^*\rangle$ by (V.11), whence ${}^t u(i(\gamma)) = h^{-1}(f_*\gamma) \in \Gamma$ and the theorem is proved.

Remarks V.14

(1) We have $\dim \mathrm{Alb}(X) = \dim H^0(X, \Omega^1_X)$. In particular, if $H^0(X, \Omega^1_X) = 0$ (for example if $X = \mathbb{P}^1$, or if X is a surface with $q = 0$), every morphism from X to a complex torus is trivial (i.e. the image is reduced to a point).

(2) From the universal property it follows immediately that the Albanese variety is functorial in nature: if $f : X \to Y$ is a morphism of smooth projective varieties, there exists a unique morphism $F : \mathrm{Alb}(X) \to \mathrm{Alb}(Y)$ such that the diagram

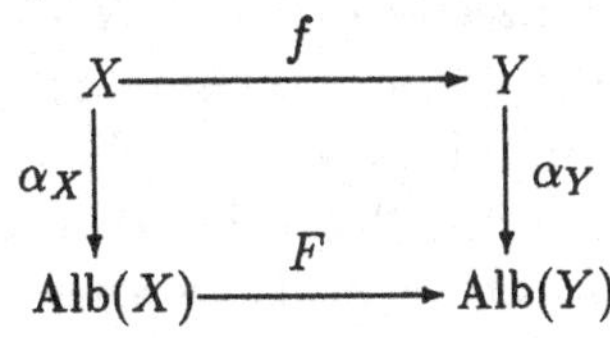

is commutative.

(3) From the universal property we also deduce that the Abelian variety $\mathrm{Alb}(X)$ is generated by $\alpha(X)$ (because the Abelian subvariety of $\mathrm{Alb}(X)$ generated by $\alpha(X)$ satisfies the universal property). In particular $\alpha(X)$ is not reduced to a point if $\mathrm{Alb}(X) \neq (0)$. It also follows that if the morphism $f : X \to Y$ is surjective, then so is the morphism $F : \mathrm{Alb}(X) \to \mathrm{Alb}(Y)$ obtained from f in (2).

(4) If X is a curve, $\mathrm{Alb}(X)$ is equal to the Jacobian JX.

(5) From the construction of $A = \mathrm{Alb}(X)$ it follows that the map $\alpha_* : H_1(X, \mathbb{Z}) \to H_1(A, \mathbb{Z})$ is surjective, and that its kernel is the torsion subgroup of $H_1(X, \mathbb{Z})$. Geometrically, this implies that the inverse image under α of a connected étale cover of A is connected.

Proposition V.15 *Let S be a surface, $\alpha : S \to \mathrm{Alb}(S)$ the Albanese map. Suppose that $\alpha(S)$ is a curve C. Then C is a smooth curve of genus q, and the fibres of α are connected.*

We sometimes call the map $\alpha : S \to C$ the Albanese fibration of S. We need the following lemma:

Lemma V.16 *Suppose that α factorizes as $S \xrightarrow{f} T \xrightarrow{j} \mathrm{Alb}(S)$, with f surjective. Then $\tilde{j} : \mathrm{Alb}(T) \to \mathrm{Alb}(S)$ is an isomorphism.*

Proof The functoriality of the Albanese variety (V.14(2)) provides a morphism $F : \mathrm{Alb}(S) \to \mathrm{Alb}(T)$. We have a commutative diagram

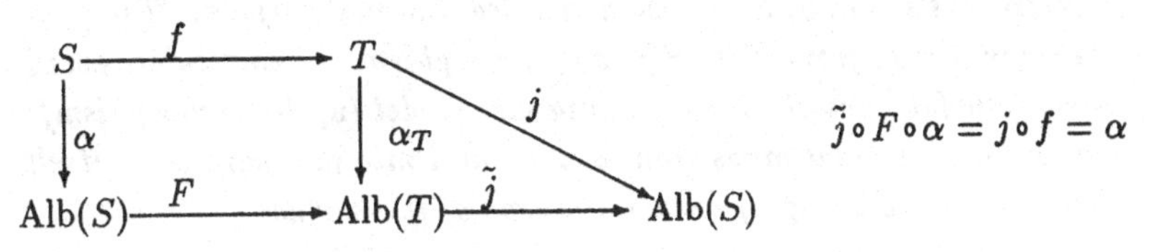

whence $\tilde{j} \circ F = \mathrm{Id}$ by the universal property of α. Since F is surjective (V.14(3)) $\tilde{j}$ and F are isomorphisms inverse to each other.

Proof of Proposition V.15 Let N be the normalization of C. Since S is normal, α factorizes as $\alpha : S \xrightarrow{f} N \xrightarrow{j} \mathrm{Alb}(S)$. By the lemma, $\tilde{j} : JN \to \mathrm{Alb}(S)$ is an isomorphism. Since $\alpha_N : N \to JN$ is an embedding, so is j, which proves that $N = C$. Thus C is a smooth curve of genus q. To prove that α has connected fibres, we 'recall' the following result, which is essentially the same as Zariski's 'connectedness theorem' (cf. [EGA III, 4.3]):

Fact V.17 (Stein factorization)
Let $f : X \to Y$ be a proper morphism (of varieties or schemes). Then f factorizes as $f : X \xrightarrow{p} \tilde{Y} \xrightarrow{g} Y$, where g is a finite morphism and p is a surjective morphism with connected fibres.

End of proof of V.15 Let us factorize α as $\alpha : S \xrightarrow{p} \tilde{C} \xrightarrow{g} C$ as above; noticing that, by replacing $\tilde{C}$ by its normalization if necessary, we may assume that $\tilde{C}$ is smooth. It follows from Lemma V.16 that g induces an isomorphism $G : J\tilde{C} \xrightarrow{\sim} JC$ such that $G \circ \alpha_{\tilde{C}} = \alpha_C \circ g$; this implies that g is an isomorphism, which completes the proof of the proposition.

Proposition V.15 will be used in an essential way in the following case:

Lemma V.18 *Let S be a surface with $p_g = 0$, $q \geqslant 1$, $\alpha : S \to \mathrm{Alb}(S)$ its Albanese map. Then $\alpha(S)$ is a curve.*

Proof If $\alpha(S)$ is a surface, the morphism $\overline{\alpha} : S \to \alpha(S)$ is generically finite, hence étale over an open subset $U \in \alpha(S)$. Let $x \in U$; $\alpha(S)$ is smooth at x, and we can take local coordinates $u_1, \ldots, u_q$ for $\mathrm{Alb}(S)$ at x such that $\alpha(S)$ is defined locally by $u_3 = \cdots = u_q = 0$. Since $A = \mathrm{Alb}(S)$ is parallelizable, there exists a 2-form $\omega \in H^0(A, \Omega_A^2)$ such that ω and $du_1 \wedge du_2$ take the same value at x; but then $\alpha^*\omega$ is a global 2-form on S, non-zero above x, a contradiction.

Theorem V.19 *Let S, S' be two non-ruled minimal surfaces. Then every birational map from S' to S is an isomorphism. In particular, every non-ruled surface admits a unique minimal model (up to isomorphism); the group of birational maps from a non-ruled minimal surface to itself coincides with the group of automorphisms of the surface.*

Proof Let ϕ be a birational transformation from S' to S. By the theorem on resolution of indeterminacy, there exists a commutative diagram

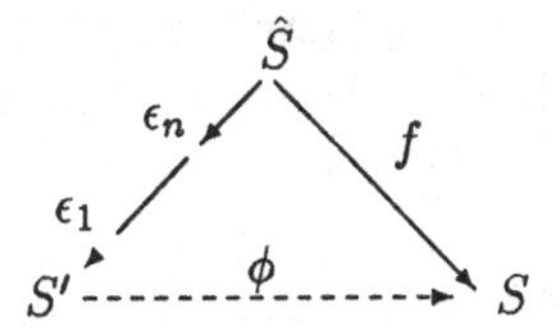

where the ϵ_i are blow-ups, and f a morphism. Among all the diagrams of this type, let us choose one with n minimal. If $n = 0$, the theorem is proved; so suppose $n \neq 0$. Let E be the exceptional curve of the blow-up ϵ_n. The image $f(E)$ is a curve C on S, otherwise f would factorize as $f' \circ \epsilon_n$, contradicting the minimality of n.

We now calculate $C.K_S$. Notice that if $\epsilon : \hat{X} \to X$ is the blow-up of a point on the surface X, and $\hat{\Gamma}$ is an irreducible curve on $\hat{X}$ such that $\epsilon(\hat{\Gamma})$ is a curve Γ, we have $K_{\hat{X}}.\hat{\Gamma} = (\epsilon^* K_X + E).(\epsilon^*\Gamma - mE)$, with $m = E.\hat{\Gamma}$ (cf. II.3) whence $K_{\hat{X}}.\hat{\Gamma} = K_X.\Gamma + m \geqslant K_X.\Gamma$, with equality only when $\hat{\Gamma}$ does not meet the exceptional divisor.

The birational morphism f being composed of blow-ups, we get $K_S.C \leqslant K_{\hat{S}}.E = -1$, with equality if and only if E does not meet any of the curves contracted by f. But in that case the restriction of f to E is an isomorphism, so that C is a rational curve with $K.C = -1$, i.e. an exceptional curve, which is impossible. Thus $K.C \leqslant -2$, so $C^2 \geqslant 0$ (by the genus formula).

Note that these two inequalities imply that all the P_n vanish: if $|nK|$ contained a divisor D (for $n \geqslant 1$), we would have $D.C \geqslant 0$ by the useful remark, thus $K.C \geqslant 0$, a contradiction. We must now distinguish two cases:

If $q = 0$, Castelnuovo's theorem says that S is rational, which is excluded.

If $q > 0$, the Albanese map of S gives a surjective morphism $p : S \to B$ with connected fibres, where B is a smooth curve of genus q (Lemma V.18 and Proposition V.15).

Since C is rational, C is contained in a fibre of p, say F; since $C^2 \geqslant 0$, Lemma III.9 shows that $F = rC$ for some integer r. This tells us that $C^2 = 0$, hence $C.K = -2$. Then by the genus formula $r = 1$, $g(F) = 0$. At this point the Noether–Enriques theorem (III.14) implies that S is ruled, a contradiction.

Historical Note V.20

Castelnuovo proved Corollary V.5 first ([C1]), then Theorem V.1 ([C2]). His proof is based on 'the termination of adjunction' (i.e. the relation $|D+nK| = \emptyset$ for all sufficiently large n, for all divisors D), like the one given here, but more complicated. The proof given here is essentially Kodaira's (cf. [S2]).

There is a proof valid in all characteristics due to Zariski ([Z2] and [Z3]).

The uniqueness of the minimal model for non-ruled surfaces appeared in 1901, in the equivalent form of 'non-existence of exceptional curves of the second kind' ([C-E]); it already depended on Castelnuovo's theorem. A modern version was given by Zariski in [Z2].

The classification of minimal rational surfaces seems to have been stated for the first time by Vaccaro ([Va]); our proof is due to Andreotti ([A]), with a slight improvement from Kodaira.

Exercises V.21

(1) Let S be a surface for which the anticanonical system $|-K|$ is ample (i.e. some multiple of it determines an embedding of S into $\mathbb{P}^N$). Show that either $S = \mathbb{P}^1 \times \mathbb{P}^1$, or S is obtained from $\mathbb{P}^2$ by blowing up r distinct points ($r \leqslant 8$) in general position (cf. Chapter IV, Exercise 11).
(Show that S is rational with the help of Castelnuovo's theorem; then use V.10, noting that if S dominates a surface $\mathbb{F}_n$ ($n \geqslant 2$), $-K_S$ is not ample.)

(2) Let S be a surface in $\mathbb{P}^n$, H a hyperplane section. Assume that $H \equiv -K$. Show that S is
either a del Pezzo surface S_d ($3 \leqslant d \leqslant 9$);
or the surface S'_8, the image of $\mathbb{P}^1 \times \mathbb{P}^1$ embedded in $\mathbb{P}^8$ by the system $|2h_1 + 2h_2|$ in the notation of I.9(b) (this surface is often counted as a del Pezzo surface).

(3) Let S be a non-ruled surface. Show that for every embedding of S in $\mathbb{P}^n$, the group of automorphisms of $\mathbb{P}^n$ fixing S is finite (using the theory of algebraic groups). Deduce that the group of automorphisms of S is an extension of a discrete group by an Abelian variety of dimension $\leqslant q$.

(4) Let G be the group of automorphisms of the surface $\mathbb{F}_n$, $n \geqslant 1$. Show that there exists an exact sequence

$$1 \to T \to G \to PGL(2, \mathbb{C}) \to 1$$

where T is the semi-direct product of $\mathbb{C}^*$ with $H^0(\mathbb{P}^1, \mathcal{O}_{\mathbb{P}^1}(n))$, where $\mathbb{C}^*$ acts by multiplication.
Calculate $\mathrm{Aut}(\mathbb{F}_0)$.

(5) Show that a surface containing an infinite number of exceptional curves is rational, and that there exist such surfaces. (Let P be a pencil of cubic curves in $\mathbb{P}^2$ such that each cubic of P is irreducible, and let S be the surface obtained by blowing up $\mathbb{P}^2$ in the base points of P. Show that any divisor D on S with $D^2 = -1$, $K.D = -1$ is equivalent to an exceptional curve. Prove that $\mathrm{Pic}\, S$ contains infinitely many such classes; for this consider the class $\delta = L - E_1 - E_2 - E_3$ in $\mathrm{Pic}\, S$ (with the notation of IV.2(4)), and the automorphism $D \mapsto D + (\delta.D)\delta$ of $\mathrm{Pic}\, S$.)

VI

SURFACES WITH $p_g = 0$ AND $q \geqslant 1$

The object of this chapter is to give a complete classification of the surfaces mentioned in the title. In particular, we get various characterizations of ruled surfaces.

Lemma VI.1

(a) *Let S be a surface with $p_g = 0$, $q \geqslant 1$. Then $K^2 \leqslant 0$, and $K^2 < 0$ unless $q = 1$ and $b_2 = 2$.*

(b) *Let S be a minimal surface with $K^2 < 0$; then $p_g = 0$ and $q \geqslant 1$.*

Proof Since $2q = b_1$ (III.19), Noether's formula gives

$$12 - 12q = K^2 + 2 - 4q + b_2 , \quad \text{or} \quad K^2 = 10 - 8q - b_2 .$$

To prove (a) we must check that $b_2 \geqslant 2$ if $q = 1$. Consider the Albanese map $\alpha : S \to B$, where $B = \mathrm{Alb}(S)$ is an elliptic curve. Let $f \in H^2(S, \mathbb{Z})$ be the class of a generic fibre of α and h that of a hyperplane section. Since $f^2 = 0$ and $h.f > 0$, h and f are linearly independent in $H^2(S, \mathbb{Z})$, and so $b_2 \geqslant 2$.

(b) Suppose $p_g \neq 0$, and let $D \in |K|$, $D = \sum n_i C_i$, $n_i > 0$. Since $K.D < 0$, $K.C_i$ must be negative for some i; since $C_i.C_j \geqslant 0$ for $i \neq j$, this implies $C_i^2 < 0$. Hence C_i is exceptional, a contradiction. The same argument gives $P_n = 0$ for all n; if $q = 0$, then S would be rational by Castelnuovo's theorem, and so $K^2 = 8$ or 9.

Proposition VI.2 *Let S be minimal with $K^2 < 0$. Then S is ruled.*

Proof By Lemmas VI.1(b) and V.18, the Albanese map $p : S \to B$ has connected fibres and B is a smooth curve. Assume that S is not ruled.

Step 1: *Let $C \subset S$ be an irreducible curve with $K.C < 0$ and $|K+C| = \emptyset$. Then $p_{|C}$ is étale, and is an isomorphism if $q \geqslant 2$* (i.e. C is a section of p). *Thus $g(C) = q$.*

Proof Apply Riemann–Roch to $K + C$:

$$0 = h^0(K + C) \geqslant \chi(\mathcal{O}_S) + \frac{1}{2}(C^2 + C.K) = 1 - q + g(C) - 1 ,$$

and so $g(C) \leqslant q$.

Since S is minimal, $C^2 \geqslant 0$, and so C cannot lie in a reducible fibre of p, by Lemma III.9. If C were a fibre of p, then $C^2 = 0$, and so $C.K = -2$ and $g(C) = 0$; but then S would be ruled, by the theorem of Noether and Enriques (III.4). Hence $p(C) = B$. Let N denote the normalization of C; p then defines a ramified cover $N \to B$, of degree d, say. By Riemann–Hurwitz, we get

$$g(N) = 1 + d(g(B) - 1) + \frac{r}{2} ,$$

where r is the number of branch points, counted with their index. Thus

$$q \geqslant g(C) \geqslant g(N) \geqslant 1 + d(q - 1) .$$

Hence either $d = 1$, or $q = 1$ and $C = N$, and the assertion follows.

Step 2: *There is an irreducible curve C on S with $|K + C| = \emptyset$ and $K.C < -1$.*

Proof By V.8, there is an effective divisor D such that $|K + D| = \emptyset$ and $K.D < -1$. Say $D = \sum_{i=1}^{r} n_i C_i$, $n_i > 0$; removing some of the C_i if necessary, we may assume that $K.C_i < 0$ for all i. We shall show that D is then in fact irreducible.

(a) Suppose that $n_i \geqslant 2$ for some i, so that $|K + 2C_i| = \emptyset$. Riemann–Roch gives

$$\begin{aligned} 0 &= h^0(2C_i + K) \geqslant 1 - q + 2C_i^2 + C_i.K \\ &= 1 - q + 2(C_i^2 + C_i.K) - C_i.K . \end{aligned}$$

By Step 1, $C_i^2 + C_i.K = 2(q - 1)$, and so $0 > 3(q - 1)$, a contradiction.

(b) Suppose that $r \geqslant 2$; then $|K + C_1 + C_2| = \emptyset$, and by Step 1 again we get

$$\begin{aligned} 0 &= h^0(K + C_1 + C_2) \\ &= h^1(K + C_1 + C_2) + 1 - q + \frac{1}{2}(C_1^2 + C_1.K) \\ &\quad + \frac{1}{2}(C_2^2 + C_2.K) + C_1.C_2 \\ &= (q - 1) + h^1(K + C_1 + C_2) + C_1.C_2 , \end{aligned}$$

which is impossible unless $C_1.C_2 = h^1(K + C_1 + C_2) = 0$. But if $C_1 \cap C_2 = \emptyset$, there is an exact sequence

$$0 \to \mathcal{O}_S(-C_1 - C_2) \to \mathcal{O}_S \to \mathcal{O}_{C_1} \oplus \mathcal{O}_{C_2} \to 0\,,$$

and the corresponding cohomology sequence gives $H^1(S, \mathcal{O}_S(-C_1 - C_2)) \neq 0$. Hence also $h^1(K + C_1 + C_2) \neq 0$, which is a contradiction.

Thus D is irreducible, and Step 2 is proved.

Step 3: *We get a contradiction.*

Let C be an irreducible curve in S with $C.K < -1$ and $|K + C| = \emptyset$. Suppose first that C is a section of p. Riemann–Roch gives

$$h^0(C) \geqslant 1 - q + \frac{1}{2}(C^2 - C.K) = -C.K \geqslant 2\,;$$

in other words, C moves in its linear equivalence class. Let F be a generic fibre of p; then the point $C \cap F$ moves linearly on F, and so F must be rational, which is a contradiction.

Suppose then that $q = 1$ and $p_{|C}$ is étale. The inclusion $i : C \hookrightarrow S$ defines a section $e : C \hookrightarrow S \times_B C$; let S' denote the connected component of $S \times_B C$ that contains $e(C) = C'$, say. The projection $\pi : S' \to S$ is étale, and so $\Omega^1_{S'} \cong \pi^*\Omega^1_S$ and $K_{S'} \equiv \pi^* K_S$. Hence

$$K_{S'}.C' = \deg_{C'}(e^* K_{S'}) = \deg_C(i^* K) = K.C < -1$$

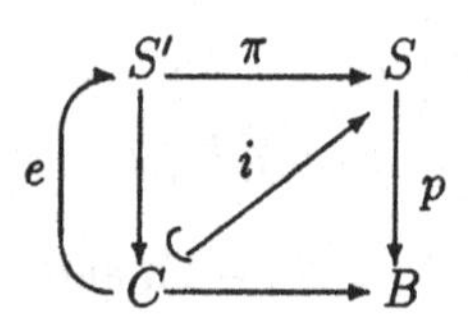

(see I.6). Hence as usual $p_g(S') = 0$, and so by Riemann–Roch

$$h^0(C') \geqslant \chi(\mathcal{O}_{S'}) - 1 + g(C') - K_{S'}.C'\,.$$

Since $\chi(\mathcal{O}_{S'}) = 0$ by Lemma VI.3 below, we get $h^0(C') \geqslant 2$, which leads to a contradiction as before.

Lemma VI.3 *Let $\pi : S' \to S$ be an étale map of surfaces, $\deg \pi = n$. Then $K^2_{S'} = n\,K^2$, $\chi_{\text{top}}(S') = n\,\chi_{\text{top}}(S)$ and $\chi(\mathcal{O}_{S'}) = n\,\chi(\mathcal{O}_S)$.*

Proof The last equation follows from the first two and Noether's formula. The first is obvious, since $K_{S'} \equiv \pi^* K$. The second is well known to topologists, and can be proved in the following way. Choose a triangulation of S; then $\chi_{\text{top}}(S) = \sum (-1)^i f_i(S)$, where $f_i(S)$ is the number of faces of dimension i. Since the faces are simply connected, their inverse images in S' triangulate it. Clearly $f_i(S') = n\, f_i(S)$, and we are done.

To complete the classification of surfaces with $p_g = 0$ and $q \geqslant 1$, we must consider the case $K^2 = 0$, $q = 1$, $b_2 = 2$ (see Lemma VI.1). We shall need the following lemmas.

Lemma VI.4 *Let S be a surface, B a smooth curve and $p : S \to B$ a surjective morphism. Let $\Sigma \subset B$ be the (finite) set of points over which p is not smooth, and let $\eta \in B - \Sigma$. Let F_b denote $p^{-1}(b)$, $b \in B$. Then*

$$\chi_{\text{top}}(S) = \chi_{\text{top}}(B)\, \chi_{\text{top}}(F_\eta) + \sum_{s \in \Sigma} (\chi_{\text{top}}(F_s) - \chi_{\text{top}}(F_\eta)) \ .$$

Proof Recall that for any topological space X and any closed subset $F \subset X$, there is a long exact sequence

$$\cdots \to H^i_c(X - F, \mathbb{Z}) \to H^i(X, \mathbb{Z}) \to H^i(F, \mathbb{Z}) \to H^{i+1}_c(X - F, \mathbb{Z}) \to \cdots$$

(where the subscript c means cohomology with compact support), and hence $\chi_{\text{top}}(X) = \chi_{\text{top}}(F) + \chi_{\text{top},c}(X - F)$.

Set $U = S - \bigcup_{s \in \Sigma} F_s$; by the above formula,

$$\chi_{\text{top}}(S) = \chi_{\text{top},c}(U) + \sum_{s \in \Sigma} \chi_{\text{top}}(F_s) \ ,$$

and also

$$\chi_{\text{top}}(B) = \chi_{\text{top},c}(B - \Sigma) + \chi(\Sigma) = \chi_{\text{top},c}(B - \Sigma) + \text{Card}\, \Sigma \ .$$

Now $p : U \to B - \Sigma$ is a topological fibre bundle, and so

$$\chi_{\text{top},c}(U) = \chi_{\text{top},c}(B - \Sigma)\, \chi_{\text{top}}(F_\eta) \ .$$

The lemma follows from these three equations.

Lemma VI.5 *Let C be a reduced (but possibly reducible) curve. Then $\chi_{\text{top}}(C) \geqslant 2\chi(\mathcal{O}_C)$; equality holds if and only if C is smooth.*

Proof Let $n : N \to C$ be the normalization of C. Consider the diagram

$$\begin{array}{ccccccccc} 0 & \longrightarrow & \mathbb{C}_C & \longrightarrow & n_*\mathbb{C}_N & \longrightarrow & \epsilon & \longrightarrow & 0 \\ & & \downarrow & & \downarrow & & \downarrow \phi & & \\ 0 & \longrightarrow & \mathcal{O}_C & \longrightarrow & n_*\mathcal{O}_N & \longrightarrow & \delta & \longrightarrow & 0\,. \end{array}$$

where $\mathbb{C}_X$ denotes the constant sheaf $\mathbb{C}$ on the variety X, and ϵ, δ are defined so as to make the rows exact.

To say that ϕ is injective is the same (diagram chasing) as saying that a local section of $n_*\mathcal{O}_N$ that comes both from $n_*\mathbb{C}_N$ and $\mathcal{O}_C$ in fact comes from $\mathbb{C}_C$; this, however, is obvious.

Thus $h^0(\delta) \geqslant h^0(\epsilon)$, and it follows from the diagram that

$$\begin{aligned} \chi_{\mathrm{top}}(N) &= \chi_{\mathrm{top}}(C) + h^0(\epsilon) \\ \chi(\mathcal{O}_N) &= \chi(\mathcal{O}_C) + h^0(\delta), \end{aligned}$$

and so $\chi_{\mathrm{top}}(C) = 2\chi(\mathcal{O}_C) + h^0(\delta) + (h^0(\delta) - h^0(\epsilon))$, since $\chi_{\mathrm{top}}(N) = 2\chi(\mathcal{O}_N)$. Hence $\chi_{\mathrm{top}}(C) \geqslant 2\chi(\mathcal{O}_C)$, and equality implies $h^0(\delta) = 0$, so that $\delta = 0$ and $C = N$.

Proposition VI.6 *Let S be a minimal surface with $p_g = 0$, $q = 1$ and $K^2 = 0$, $p : S \to B$ the Albanese map (B is an elliptic curve) and g the genus of a generic fibre of p. Then if $g \geqslant 2$, p is smooth, and if $g = 1$, then the singular fibres of p are of the form $F_b = nE$, where E is a smooth elliptic curve.*

Proof We shall show first that the fact that $b_2 = 2$ (proved in (VI.1(a)) implies that p has irreducible fibres. Suppose that some fibre contains two irreducible components F_1 and F_2; let H be a hyperplane section of S. It is enough to show that F_1, F_2, H are linearly independent in $H^2(S,\mathbb{Z})$; suppose then that $\alpha H + \beta F_1 + \gamma F_2 = 0$ in $H^2(S,\mathbb{Z})$. Let F be a generic fibre. If $\alpha \neq 0$, then $H.F = 0$, which is absurd, and so $\alpha = 0$. Thus $F_2 = r.F_1$, $r \in \mathbb{Q}$; intersecting with H gives $r > 0$, and now intersecting with F_1 (using Lemma III.9) gives $r < 0$, which is a contradiction.

Thus p has irreducible fibres; they might, however, be multiple, i.e. $F_s = nC$ for some irreducible curve C, $n \geqslant 1$. In this case

$$\begin{aligned} \chi_{\mathrm{top}}(F_s) &= \chi_{\mathrm{top}}(C) \geqslant 2\chi(\mathcal{O}_C) \qquad \text{by Lemma VI.5,} \\ 2\chi(\mathcal{O}_C) &= -C^2 - C.K = -\frac{1}{n}F_s.K = -\frac{1}{n}F_\eta.K = \frac{2}{n}\chi(\mathcal{O}_{F_\eta}) \\ &= \frac{1}{n}\chi_{\mathrm{top}}(F_\eta)\,. \end{aligned}$$

Since $g \geqslant 1$, i.e. $\chi_{\text{top}}(F_\eta) \leqslant 0$, we get finally $\chi_{\text{top}}(F_s) \geqslant \chi_{\text{top}}(F_\eta)$, and equality holds if and only if both $\chi_{\text{top}}(C) = 2\chi(\mathcal{O}_C)$ (i.e. C is smooth) and $\frac{1}{n}\chi_{\text{top}}(F_\eta) = \chi_{\text{top}}(F_\eta)$ (i.e. either $n = 1$ or $g = 1 = g(C)$).

Hence for $s \in \Sigma$ (in the notation of Lemma VI.4), $\chi_{\text{top}}(F_s) - \chi_{\text{top}}(F_\eta) \geqslant 0$, and equality holds only if $F_s = nE$, E a smooth elliptic curve, $g(F_\eta) = 1$. Now apply the formula of VI.4; we get $\chi_{\text{top}}(S) = 2 - 2b_1 + b_2 = 0$ and $\chi_{\text{top}}(B) = 0$, and so $\sum_{s\in\Sigma}(\chi_{\text{top}}(F_s) - \chi_{\text{top}}(F_\eta)) = 0$. The proposition follows, by what we already know.

Smooth fibrations $S \to B$ are very special and can be completely classified. If there are multiple fibres ($g = 1$), we can reduce to a smooth fibration by the following result:

Lemma VI.7 *Let $p : S \to B$ be a morphism from a surface onto a smooth curve whose fibres are either smooth or multiples of smooth curves. Then there is a ramified Galois cover $q : B' \to B$ with Galois group G, say, a surface S' and a commutative diagram*

$$\begin{array}{ccc} S' & \xrightarrow{q'} & S \\ {\scriptstyle p'}\downarrow & & \downarrow{\scriptstyle p'} \\ B' & \xrightarrow{q} & B \end{array}$$

such that the action of G on B' lifts to S', q' induces an isomorphism $S'/G \xrightarrow{\sim} S$ and p' is smooth.

Proof It is enough to eliminate each multiple fibre by taking successive branched covers, and so the lemma follows from the following local version.

Lemma VI.7′ *Let $\Delta \subset \mathbb{C}$ be the unit disc, U a (non-compact) smooth analytic surface and $p : U \to \Delta$ a morphism that is smooth over $\Delta - \{0\}$, such that $p^*0 = nC$ for some smooth curve $C \subset U$. Let $q : \Delta \to \Delta$ be the morphism defined by $z \mapsto z^n$, $\tilde{U} = U \times_\Delta \Delta$, U' the normalization of $\tilde{U}$ and p', q' the projections of U' onto Δ, U. The group μ_n of nth roots of unity acts on Δ (by $z \mapsto \zeta z$), and so on $\tilde{U}$ (via the second factor), and so on U'; q' induces an isomorphism $U'/\mu_n \xrightarrow{\sim} U$. The fibration $p' : U' \to \Delta$ is smooth.*

Proof Since the hypotheses and statements are local on U, we may assume that there are local coordinates x, y on U such that $p(x, y) = x^n$, and C is defined by $x = 0$. Then

$$\begin{aligned}\tilde{U} &= \{(x,y,z) \mid (x,y) \in U \ , \ z \in \Delta \text{ and } x^n = z^n\} \\ &= \bigcup_{\zeta \in \mu_n} U_\zeta, \text{ where } U_\zeta = \{(x,y,\zeta x) \mid (x,y) \in U\} \ .\end{aligned}$$

Each U_ζ is isomorphic to U, via $\tilde{q} : \tilde{U} \to U$; $\tilde{U}$ is the union of the n varieties U_ζ, identified along the line $x = 0$. U' is then the disjoint union of the U_ζ and μ_n acts on U' by interchanging the components. Identifying U_ζ with U expresses $p' : U_\zeta \to \Delta$ as $(x,y) \mapsto \zeta x$. All the statements in the lemma are now clear.

Note that the construction described above does not tell us that the cover $q : B' \to B$ is ramified only over those points of B that correspond to multiple fibres, nor that $q' : S' \to S$ is étale. That in fact there is such a q follows from the arguments below.

Proposition VI.8 *Let $p : S \to B$ be a smooth morphism from a surface to a curve, and F a fibre of p. Assume either that $g(B) = 1$ and $g(F) \geqslant 1$, or that $g(F) = 1$. Then there exists an étale cover B' such that the fibration $p' : S' = S \times_B B' \to B'$ is trivial, i.e. $S' \cong B' \times F$. Furthermore, we can take the cover $B' \to B$ to be Galois with group G, say, so that $S \cong (B' \times F)/G$.*

Proof This depends upon a series of facts from the theory of moduli spaces for curves; a good reference is [Gr].

Let T be a variety. A curve of genus g over T is a smooth morphism $f : X \to T$ whose fibres are curves of genus g. f is a topological fibre bundle, and so the sheaf $R^1 f_*(\mathbb{Z}/n\mathbb{Z})$ is locally constant for all n. There is a symplectic form on it given by the cup product:

$$R^1 f_*(\mathbb{Z}/n\mathbb{Z}) \otimes R^1 f_*(\mathbb{Z}/n\mathbb{Z}) \to (\mathbb{Z}/n\mathbb{Z})_T \ .$$

Having such a sheaf is equivalent to knowing its fibre at a point $t \in T$ (i.e. $H^1(X_t, \mathbb{Z}/n\mathbb{Z})$) together with the action of the fundamental group $\pi_1(T,t)$: this action preserves the symplectic form on $H^1(X_t, \mathbb{Z}/n\mathbb{Z})$.

Let us endow the constant sheaf $(\mathbb{Z}/n\mathbb{Z})_T^{2g}$ with its standard symplectic form. A J_n-rigidified curve of genus g over T is a curve of genus g over T together with a symplectic isomorphism $(\mathbb{Z}/n\mathbb{Z})_T^{2g} \xrightarrow{\sim} R^1 f_*(\mathbb{Z}/n\mathbb{Z})$. A curve of genus g over T can be J_n-rigidified if $\pi_1(T,t)$ acts trivially on $H^1(X_t, \mathbb{Z}/n\mathbb{Z})$; this can certainly fail to hold, but since $\mathrm{Aut}(H^1(X_t, \mathbb{Z}/n\mathbb{Z}))$ is finite, π_1 has a subgroup of finite index which acts trivially, and so there is an étale cover $T' \to T$ such that the pullback of the given curve is J_n-rigidifiable over T'.

Let $n \geqslant 3$. One then shows that J_n-rigidification eliminates automorphisms, and this implies that there is a universal J_n-rigidified curve of genus g, denoted by $U_{g,n} \to T_{g,n}$. In other words, every J_n-rigidified curve of genus g over a variety T is the pullback from $U_{g,n} \to T_{g,n}$ via a unique morphism $T \to T_{g,n}$. The spaces $R_{g,n}$ are in fact quasi-projective varieties ([M2]), but it is enough for our purposes that they exist as analytic spaces ([Gr]). We shall need also the following properties:

(1) For $g \geqslant 2$, there is no non-constant analytic morphism $h : \mathbb{C} \to T_{g,n}$.

(2) For $g = 1$, there is no non-constant analytic morphism from a connected compact variety X to $T_{1,n}$.

(2) is elementary, since the j-invariant defines a holomorphic function on X, which must be constant.

(1) is more subtle. One can use the hard fact that the universal cover T_g (Teichmüller space) of $T_{g,n}$ is a bounded domain, and so cannot be the target of any non-trivial morphism from $\mathbb{C}$. One can also consider the space $A_{g,n}$, which classifies J_n-rigidified principally polarized Abelian varieties of dimension g; by construction, its universal cover is the Siegel upper half-space H_g, which is a bounded domain. Finally one applies the Torelli theorem, which shows that the map $T_{g,n} \to A_{g,n}$ obtained by sending a curve to its Jacobian is finite.

We shall show how the proposition follows from properties (1) and (2). Let $p : S \to B$ be a smooth morphism, with fibres of genus g; then for fixed $n \geqslant 3$, there is an étale cover $B' \to B$ such that the curve $S' \to B'$ (with $S' := S \times_B B'$) is J_n-rigidifiable. Choose some J_n-rigidification; we get a morphism $h : B' \to T_{g,n}$ such that $S' = U_{g,n} \times_{T_{g,n}} B'$. If $g(B) = 1$, then $g(B') = 1$; if $g \geqslant 2$, then h is trivial by (1), since the universal cover of B' is $\mathbb{C}$. If $g = 1$, then h is trivial by (2). The proposition follows.

Corollary VI.9 *Let S be a minimal non-ruled surface with $p_g = 0$, $q = 1$ and $K^2 = 0$. Then there are two curves B, F, of genus $\geqslant 1$, and a finite group G of automorphisms of B acting on $B \times F$ compatibly with its action on B (i.e. $g(b, f) = (gb, \cdot)$ for $g \in G$, $b \in B$, $f \in F$) such that $S \cong (B \times F)/G$. The curve B/G is elliptic; if moreover $g(F) \geqslant 2$, then B is elliptic and G is a group of translations of B.*

This follows at once from VI.6–8.

Lemma VI.10 *Let B, F be curves of genus $\geqslant 1$ and G a group of automorphisms of B acting on $B \times F$ compatibly with its action on B.*

(i) *If $g(F) \geqslant 2$, then G acts on F and $g(b,f) = (gb, gf)$ for $g \in G$, $b \in B$, $f \in F$.*

(ii) *If $g(F) = 1$, there is an étale cover $\tilde{B}$ of B and a group H acting on $\tilde{B}$ and F such that $\tilde{B}/H \cong B/G$ and $(\tilde{B} \times F)/H \cong (B \times F)/G$.*

Proof

Step 1: For $g \in G$ and $b \in B$, we have $g(b,f) = (gb, \phi_g(b).f)$, where $\phi_g(b)$ is an automorphism of F depending continuously on B. If $g(F) \geqslant 2$, then $\mathrm{Aut}(F)$ is finite, and so $\phi_g(b)$ is independent of b; this proves (i).

So suppose that F is elliptic; fix an origin 0 on F, so that it has a natural group structure (i.e. it is an Abelian variety). Then $\phi_g(b)$ is of the form $f \mapsto a_g(b).f + t_g(b)$, where $a_g(b)$ is an automorphism of F as an Abelian variety and $t_g(b)$ is a translation (see Proposition V.12). The group $\mathrm{Aut}_0(F)$ of automorphisms of F that preserve the group structure is finite (V.12 or VI.16 below), so that $a_g(b) = a_g$ is independent of b, and one can check easily that $t_g : B \to F$ is a morphism. For $g, h \in G$, we have $a_{gh} = a_g a_h$, and so $a : G \to \mathrm{Aut}_0(F)$ is a group homomorphism.

Step 2: *There is a morphism $\rho : B \to F$ and an integer n such that*

$$\rho(gb) - a_g.\rho(b) = n\, t_g(b) \quad \text{for } b \in B, g \in G\,. \qquad (*)$$

Proof Recall that there is a canonical isomorphism $F \to \mathrm{Pic}^0(F)$: $f \mapsto [f] - [0]$ (in what follows we shall let $f_1 + f_2$ denote the sum of f_1 and f_2 with respect to the group law on F, and $[f_1] + [f_2]$ the corresponding divisor of degree 2 on F). Then

$$[f_1] + \cdots + [f_r] \equiv (r-1)[0] + [\Sigma f_i] \quad \text{for } f_1, \ldots, f_r \in F\,.$$

Let $u \in \mathrm{Aut}\, F$, given by $u(f) = a(f) + t$, $a \in \mathrm{Aut}_0(F)$ and $t \in F$; if $D = (n-1)[0] + f$, then

$$\begin{aligned} u^* D &= (n-1)[u^{-1}(0)] + [u^{-1}(f)] \\ &= (n-1)[-a^{-1}t] + [a^{-1}f - a^{-1}t] \\ &\equiv (n-1)[0] + [a^{-1}f - na^{-1}t]. \end{aligned}$$

Let H be a hyperplane section of $B \times F$; the bundle $\mathcal{L} = \mathcal{O}_{B\times F}(\sum_{g\in G} g^* H)$ is G-invariant. For $b \in B$, set $\mathcal{L}_b = \mathcal{L} \otimes \mathcal{O}_{\{b\}\times F}$; this is a line bundle on F of degree $n > 0$. That $\mathcal{L}$ is G-invariant means that

$$\mathcal{L}_b = (g^*\mathcal{L}) \otimes \mathcal{O}_{\{b\}\times F} = \phi_g(b)^* \mathcal{L}_{gb}\,.$$

Define ρ by $\mathcal{L}_b = \mathcal{O}_F((n-1)[0] + [\rho(b)])$. The last equation becomes

$$\rho(b) = a_g^{-1}.\rho(gb) - na_g^{-1}.t_g(b) \ ,$$

and we are done.

Step 3: Suppose for the moment that $n = 1$; define $u \in \operatorname{Aut}(B \times F)$ by $u(b, f) = (b, f - \rho(b))$. The relation $(*)$ yields $ugu^{-1}(b, f) = (gb, a_g.f)$; in other words, u defines an isomorphism $(B \times F)/G \xrightarrow{\sim} (B \times F)/H$, where $H = uGu^{-1}$ acts on $B \times F$ by means of its action on the two factors. This proves (ii) in this case.

In general, consider the (possibly disconnected) étale cover $\pi : \tilde{B} \to B$ induced by the pullback

$$\begin{array}{ccc} \tilde{B} & \xrightarrow{\tilde{\rho}} & F \\ {\scriptstyle\pi}\downarrow & & \downarrow{\scriptstyle n} \\ B & \xrightarrow{\rho} & F \,. \end{array}$$

One has $\tilde{B} \subset B \times F$, and by $(*)$, $\tilde{B}$ is G-stable. On the other hand, the group F_n of points of order n in F also acts on $\tilde{B}$, by $\epsilon(b, f) = (b, f + \epsilon)$. Set $H = \langle G, F_n \rangle \subset \operatorname{Aut}(\tilde{B})$; one has $g\epsilon g^{-1} = a_g.\epsilon$ for $g \in G$, $\epsilon \in F_n$, and so H is a semi-direct product of G by F_n: there is a split exact sequence $1 \to F_n \to H \xrightarrow{v} G \to 1$.

Make H act on $\tilde{B} \times F$ by $h(\tilde{b}, f) = (h\tilde{b}, \phi_{vh}(\pi\tilde{b}).f)$. Since $\tilde{B}/F_n \cong B$, we have $\tilde{B}/H \cong B/G$ and $(\tilde{B} \times F)/H \cong (B \times F)/G$. Consider $u \in \operatorname{Aut}(\tilde{B} \times F)$ defined by $u(\tilde{b}, f) = (\tilde{b}.f - \tilde{\rho}(\tilde{b}))$; then

$$uhu^{-1}(\tilde{b}, f) = (h\tilde{b}, a_{vh}.f + \theta_h(\tilde{b}))$$

where $\theta_h(\tilde{b}) = a_{vh}.\tilde{\rho}(\tilde{b}) - \tilde{\rho}(h\tilde{b}) + t_{vh}(\pi\tilde{b})$. Now by $(*)$, $n.\theta_h(\tilde{b}) = 0$; i.e. $\theta_h(\tilde{b}) \in F_n$. Thus θ_h is independent of $\tilde{b}$, and the action of H on $\tilde{B} \times F$, after shifting by u, is of the required form.

If $\tilde{B}$ is not connected, let B_0 be some connected component and H_0 the subgroup of H preserving B_0; then clearly $(B_0 \times F)/H_0 \cong (\tilde{B} \times F)/H$, and the lemma is proved.

Thus in every case we get $S \cong (B \times F)/G$, where $G \subset \operatorname{Aut} B$ acts on F. Note that if a normal subgroup H of G acts trivially on F, then $S \cong (B' \times F)/G'$, where $B' = B/H$, $G' = G/H$. So henceforth we shall assume that $G \subset \operatorname{Aut} F$. It follows that G has only finitely many fixed points. However, the quotient of a smooth variety by a finite group G can only be smooth if the subvarieties fixed by an element $g \in G$ are divisors for all $g \neq 1$. (This follows from Zariski's theorem on the purity of the branch locus, which is easy to prove in this case: if $f : X \to Y$ is a finite morphism of smooth varieties, then the branch locus of f is defined by the vanishing of $\det(df)$, and so is a divisor.) Hence G acts freely on $B \times F$ (i.e. for every $g \in G - \{1\}$ and $s \in B \times F$, $gs \neq s$).

It remains to check which surfaces of the form $(B \times F)/G$ have $p_g = 0$, $q \geqslant 1$. For this we must compute the numerical invariants of a quotient variety.

Lemma VI.11 *Let X be a smooth variety and G a finite subgroup of* $\operatorname{Aut} X$. *Let $\pi : X \to Y = X/G$ denote the natural projection, and assume that Y is smooth. Then the G-invariant k-fold p-forms $\alpha \in H^0(X, (\Omega^p_X)^{\otimes k})$ are the forms $\pi^*\omega$, where ω is a k-fold rational p-form on Y such that $\pi^*\omega$ is regular on X.*

Proof We shall consider only the 1-forms; the general case follows immediately. For $V = X$ or Y, let K_V denote the function field of V and $M\Omega^1_V$ the space of rational 1-forms on V, an n-dimensional K_V-vector space (where $\dim V = n$). We must show that $\pi^* : M\Omega^1_Y \to (M\Omega^1_X)^G$ is an isomorphism. Let $\{dy_1, \dots, dy_n\}$ be a K_Y-basis of $M\Omega^1_Y$; then $\{\pi^*dy_1, \dots, \pi^*dy_n\}$ is a K_X-basis of $M\Omega^1_X$. A rational 1-form $\alpha = \sum A_i\, \pi^*dy_i$ ($A_i \in K_X$) on X is G-invariant if and only if each A_i is so, i.e. if and only if $A_i = \pi^*B_i$, for some $B_i \in K_Y$, for all i. Then $\alpha = \pi^*\omega$, where $\omega = \sum B_i\, dy_i$, and the lemma is proved.

Examples VI.12

(1) π is étale (i.e. G acts freely).

Then a form α is regular if and only if $\pi^*\alpha$ is so; therefore $\pi^* : H^0(Y, (\Omega^p_Y)^{\otimes k}) \to H^0(X, (\Omega^p_X)^{\otimes k})^G$ is an isomorphism.

(2) Curves.

Set $\Omega^1 = \omega$. We want to find the rational k-fold 1-forms α on Y such that $\pi^*\alpha \in H^0(X, \omega_X^{\otimes k})$. Where π is étale, it is necessary and sufficient that α be regular. Let $P \in Y$ be a branch point of π. G acts transitively on $\pi^{-1}(P) = \{Q_1, \dots, Q_s\}$, so that the stabilizers of the Q_i are conjugate in G; their order e_P, or e, is the ramification index of Q_i, $i = 1, \dots, s$, and $es = \operatorname{Card} G = \deg \pi$. There are local coordinates y on Y at P and x_i on X at Q_i such that $\pi^*y = x_i^e$ near Q_i. Near P, we can write $\alpha = Ay^{-r}\,(dy)^{\otimes k}$, where $A(P) \neq 0$ and $r \in \mathbb{Z}$. At Q_i we have

$$\begin{aligned} \pi^*\alpha &= \pi^*A\, x_i^{-re}(ex_i^{e-1}\, dx_i)^{\otimes k} \\ &= A_i\, x_i^{-re+k(e-1)}\,(dx_i)^{\otimes k}, \qquad \text{with } A_i(Q_i) \neq 0\,. \end{aligned}$$

Thus $\pi^*\alpha$ is regular if and only if $-re + k(e-1) \geqslant 0$, and so

$$\pi^* : H^0\left(Y, \omega_Y^{\otimes k}\left(\sum_{P \in Y} P\left[k\left(1-\frac{1}{e_P}\right)\right]\right)\right) \longrightarrow H^0(X, \omega_X^{\otimes k})^G$$

is an isomorphism ($[x]$ denotes the integral part of x; the sum is in fact only over a finite set, since $e_P = 1$ if $P \in Y$ is not a branch point).

Note that for $k = 1$, $H^0(X, \omega_X)^G \cong H^0(Y, \omega_Y)$, but the corresponding formula can fail as soon as $k \geqslant 2$.

(3) p-forms.

It is in fact true in every dimension that

$$\pi^* : H^0(Y, \Omega_Y^p) \to H^0(X, \Omega_X^p)^G$$

is an isomorphism. For this, one uses the existence of a trace map $\mathrm{Tr} : H^0(X, \Omega_X^p) \to H^0(Y, \Omega_Y^p)$ such that $\pi^* \mathrm{Tr}(\omega) = \sum_{g \in G} g^*\omega$ (see for example [G]). There is no corresponding result for k-fold forms, as we saw when X, Y were curves.

We shall not need this result.

Theorem VI.13 *Let S be a minimal non-ruled surface with $p_g = 0$, $q \geqslant 1$. Then $S \cong (B \times F)/G$, where B, F are smooth irrational curves, G is a finite group acting faithfully on B and F, B/G is elliptic, F/G is rational and one of the following conditions holds:*

I: *B is elliptic, and so G is a group of translations of B;*
II: *F is elliptic, and G acts freely on $B \times F$.*

Conversely, every surface with these properties is minimal with $p_g = 0$, $q = 1$, $K^2 = 0$ and is non-ruled.

Proof Let S be a minimal non-ruled surface with $p_g = 0$ and $q \geqslant 1$; then $K^2 = 0$ and $q = 1$ (VI.1(a) and VI.2). By VI.9 and VI.10 we know that $S \cong (B \times F)/G$, where G acts on B and F such that B/G is elliptic. Moreover, either B is elliptic (Case I) or F is elliptic (Case II). In each case G acts freely on $B \times F$, so that the projection $\pi : B \times F \to S$ is étale.

We want to know when $(B \times F)/G$ has $p_g = 0$ and $q = 1$. Note that it is minimal and non-ruled: if it contained a rational curve C, then $\pi^{-1}(C)$ would be a union of rational curves, each of which would have to map surjectively to either B or F. This is clearly impossible.

Set $\tilde{S} = B \times F$. By III.22 we have

$$H^0(\tilde{S}, \Omega^1_{\tilde{S}}) \cong H^0(B, \omega_B) \oplus H^0(F, \omega_F) ,$$

so tha

$$q(\tilde{S}) = g(B) + g(F) ,$$

and

$$H^0(\tilde{S}, \Omega^2_{\tilde{S}}) \cong H^0(B, \omega_B) \otimes H^0(F, \omega_F) ,$$

so that

$$p_g(\tilde{S}) = g(B)\, g(F) .$$

In particular, $\chi(\mathcal{O}_{\tilde{S}}) = \chi(\mathcal{O}_B)\,\chi(\mathcal{O}_F)$ so that $\chi(\mathcal{O}_{\tilde{S}}) = 0$, since B or F is elliptic. If B (resp. F) is elliptic, then $\Omega^2_{\tilde{S}} \cong q^*\omega_F$ (resp. $\Omega^2_{\tilde{S}} \cong p^*\omega_B$), where q (resp. p) denotes the projection $\tilde{S} \to F$ (resp. $\tilde{S} \to B$). Hence $K^2_{\tilde{S}} = 0$, and so by VI.3 $\chi(\mathcal{O}_S) = 0$ and $K^2_S = 0$. Since π is étale,

$$\begin{aligned} H^0(S, \Omega^1_S) &\cong H^0(\tilde{S}, \Omega^1_{\tilde{S}})^G \\ &\cong H^0(B, \omega_B)^G \oplus H^0(F, \omega_F)^G \\ &\cong H^0(B/G, \omega_{B/G}) \oplus H^0(F/G, \omega_{F/G}) , \end{aligned}$$

by examples VI.12(1) and VI.12(2). Since B/G is elliptic, $q(S) = 1$ if and only if F/G is rational; in this case, $p_g(S) = 0$, since $\chi(\mathcal{O}_S) = 0$. The theorem is proved.

Examples VI.14

(1) Constructing all the examples in Case I is straightforward. Take an elliptic curve B; G must be of the form $\mathbb{Z}/a\mathbb{Z} \oplus \mathbb{Z}/\mathbb{Z}b$ with $a, b \geqslant 1$. Choose an irrational Galois cover F of $\mathbb{P}^1$ with this group, and set $S = (B \times F)/G$.

(2) Giving an example of Case II (which does not fall into Case I) is harder, since the requirement that G act freely is fairly restrictive (for example, G cannot be cyclic). We give an example with $G = \mathbb{Z}/2\mathbb{Z} \oplus \mathbb{Z}/2\mathbb{Z}$ acting on the elliptic curve F via the automorphisms $x \mapsto -x$ and $x \mapsto x + \epsilon$ (ϵ being a point of order 2). Choose an elliptic curve B'; we seek a ramified Galois cover B of B' with group $\mathbb{Z}/2\mathbb{Z} \oplus \mathbb{Z}/2\mathbb{Z}$ such that two of the non-trivial elements of the group act on B without fixed points (we shall then adjust things so that these elements correspond to the above automorphisms of F). For this we choose an étale (resp. ramified)

double cover B_1 (resp. B_2) of B' and set $B = B_1 \times_{B'} B_2$. Let σ_1, σ_2 denote the involutions on B_1, B_2, and $\overline{\sigma}_1$, $\overline{\sigma}_2$ the involutions $(\sigma_1, 1)$, $(1, \sigma_2)$ on B. Then $G = \mathbb{Z}/2\mathbb{Z} \oplus \mathbb{Z}/2\mathbb{Z}$ is the group of the Galois cover $B \to B'$; $\overline{\sigma}_1$ and $\overline{\sigma}_1\overline{\sigma}_2$ have no fixed points, and so G acts freely on $B \times F$.

We next calculate the plurigenera P_n of the surfaces $(B \times F)/G$ so as to distinguish them numerically from elliptic ruled surfaces.

Proposition VI.15 *Let $S = (B \times F)/G$ be a surface satisfying the conditions of Theorem* VI.13.

(1) $P_4 \neq 0$ *or* $P_6 \neq 0$*; in particular* $P_{12} \neq 0$.
(2) *If B or F is not elliptic, then there is an infinite increasing sequence $\{n_i\}$ of integers such that the sequence $\{P_{n_i}\}$ tends to infinity.*
(3) *If B and F are elliptic, then* $4K \equiv 0$ *or* $6K \equiv 0$*; in particular,* $12K \equiv 0$.

Proof Note that (3) follows from (1), for if B and F are elliptic then $K_{B\times F}$ is trivial, so if $D \in |4K_S|$ (resp. $|6K_S|$), then $\pi^*D = 0$, and so $D = 0$.

We divide into cases I and II as before.

Case I
One has $H^0(S, \Omega_S^2)^{\otimes k} \cong H^0(\tilde{S}, \Omega_{\tilde{S}}^2)^{\otimes k})^G \cong [H^0(B, \omega_B^{\otimes k}) \otimes H^0(F, \omega_F^{\otimes k})]^G$. Note that $H^0(B, \omega_B^{\otimes k})$ is G-invariant, since a non-zero regular 1-form on B is translation-invariant. From example VI.12(2) we get

$$P_k(S) = \dim H^0(F, \omega_F^{\otimes k})^G = \dim H^0(F/G, \mathcal{L}_k)\ ,$$

where $\mathcal{L}_k = \omega_{F/G}^{\otimes k}(\sum_{P \in F/G} P\,[k(1 - \frac{1}{e_P})])$.
Since $F/G \cong \mathbb{P}^1$, $\mathcal{L}_k$ is determined by its degree

$$\deg \mathcal{L}_k = -2k + \sum_P \left[k\left(1 - \frac{1}{e_P}\right)\right]\ .$$

The Riemann–Hurwitz formula yields

$$2g(F) - 2 = -2n + \sum_P n\left(1 - \frac{1}{e_P}\right), \qquad \text{(R–H)}$$

cf. Example VI.12(2).

Note that if there are r ramification points, then

$$\deg \mathcal{L}_k \geqslant -2k + \sum_P \left(k\left(1 - \frac{1}{e_P}\right) - 1\right) = k\frac{2g(F)-2}{n} - r\,;$$

hence if $g(F) \geqslant 2$, $P_k(S) = \max\{\deg \mathcal{L}_k + 1, 0\} \to \infty$ as $k \to \infty$. So (2) holds in this case.

Write the ramification indices in increasing order: $e_1 \leqslant \cdots \leqslant e_r$. Formula (R–H) gives $\sum(1 - \frac{1}{e_i}) \geqslant 2$. We must show that $\deg \mathcal{L}_k \geqslant 0$ for some suitable k (dividing 12). We divide further into subcases:

(a) $r \geqslant 4$. Since $2(1 - \frac{1}{e_i}) \geqslant 1$, $\deg \mathcal{L}_2 \geqslant 0$.
By (R–H), $r \geqslant 3$, and so we can assume that $r = 3$. Then $1/e_1 + 1/e_2 + 1/e_3 \leqslant 1$.

(b) $e_1 \geqslant 3$. Then $3(1 - \frac{1}{e_i}) \geqslant 2$, and so $\deg \mathcal{L}_3 \geqslant 0$. So assume $e_1 = 2$, and $1/e_2 + 1/e_3 \leqslant \frac{1}{2}$.

(c) $e_2 \geqslant 4$. Then $\deg \mathcal{L}_4 \geqslant 0$.

(d) $e_2 = 3$. Then $e_3 \geqslant 6$, and so $\deg \mathcal{L}_6 \geqslant 0$. Thus (1) is proved in case I.

Case II

Recall the following well-known facts, which follow easily from Proposition V.12:

Theorem VI.16 *Let F be an elliptic curve, with a group structure. Every automorphism of F is the composite of a translation and a group automorphism. The non-trivial group automorphisms are the symmetry $x \mapsto -x$ and also:*

for the curve $F_i = \mathbb{C}/(\mathbb{Z} \oplus \mathbb{Z}i)$, $x \mapsto \pm ix$;

for $F_\rho = \mathbb{C}/(\mathbb{Z} \oplus \mathbb{Z}\rho)$ $(\rho^3 = 1, \rho \neq 1)$, $x \mapsto \pm\rho x$ and $x \mapsto \pm\rho^2 x$.

So let $S = (B \times F)/G$ be a surface satisfying the conditions of II. We have

$$H^0(S, (\Omega_S^2)^{\otimes k}) \cong [H^0(B, \omega_B^{\otimes k}) \otimes H^0(F, \omega_F^{\otimes k})]^G.$$

Assume that $F \neq F_i$, F_ρ (resp. $F = F_i$, resp. $F = F_\rho$). By VI.16, if ω is a non-zero regular 1-form on F, then $\omega^{\otimes 2}$ (resp. $\omega^{\otimes 4}$, resp. $\omega^{\otimes 6}$) is invariant under $\operatorname{Aut} F$. So if k is even (resp. $4|k$, resp. $6|k$), then

$$P_k(S) = \dim H^0(B, \omega_B^{\otimes k})^G = h^0\left(B/G, \mathcal{O}_{B/G}\left(\sum_P P\left[k\left(1 - \frac{1}{e_P}\right)\right]\right)\right)$$

(VI.12(2)), since B/G is an elliptic curve. Clearly this expression is non-zero, and tends to infinity as $k \to \infty$ if $g(B) \geqslant 2$ (since then there is at least one branch point). So (1) and (2) hold in case II also.

Theorem VI.17 (Enriques) *Let S be a surface with $P_4 = P_6 = 0$ (or $P_{12} = 0$). Then S is ruled.*

Proof If $q = 0$, then we are done by Castelnuovo's criterion for rationality. If $q \geqslant 1$, the result follows from VI.13 and VI.15.

Corollary VI.18 *The following conditions on a surface S are equivalent:*

(1) *S is ruled;*
(2) *there is a non-exceptional curve C on S such that $K.C < 0$;*
(3) *for every divisor D on S, $|D + nK| = \emptyset$ for $n \gg 0$ ('adjunction terminates');*
(4) *$P_n = 0$ for all n;*
(5) *$P_{12} = 0$.*

Proof (2) $\Rightarrow$ (3): Since C is not exceptional, the genus formula gives $C^2 \geqslant 0$. Since $(D + nK).C < 0$ for all $n \gg 0$, (3) follows from III.5.

(3) $\Rightarrow$ (4) $\Rightarrow$ (5) is clear; (5) $\Rightarrow$ (1) follows from VI.17.

(1) $\Rightarrow$ (2): By the structure theorem for minimal models of ruled surfaces, there is a birational morphism $f : S \to M$, where M is a geometrically ruled surface (resp. $\mathbb{P}^2$). There is a fibre F of M (resp. a line L in $\mathbb{P}^2$) over which f is an isomorphism; then $f^*F.K_S = F.K_M = -2$ (resp. $f^*L.K_S = L.K_M = -3$), and we are done.

To end this chapter we examine surfaces S of the form $(B \times F)/G$ for elliptic curves B, F (so that $12K_S \equiv 0$, by VI.15).

Definition VI.19 *A surface S is bielliptic if $S \cong (E \times F)/G$, where E, F are elliptic curves and G is a finite group of translations of E acting on F such that $F/G \cong \mathbb{P}^1$.*

The term 'hyperelliptic surfaces' is the one which appears in the current literature; but this creates confusion with the classical terminology, and seems inadequate.

We can now draw up a complete list of bielliptic surfaces. Since G is a subgroup of $\operatorname{Aut} F$, it is a semi-direct product $T \rtimes A$, where T is a group of translations and $A \subset \operatorname{Aut} F$ is a subgroup preserving the group structure. Because $F/G \cong \mathbb{P}^1$, A is non-zero and therefore isomorphic to $\mathbb{Z}/a\mathbb{Z}$, where $a = 2, 3, 4$ or 6 (VI.16).

Since G is also a group of translations of E, the product $T \rtimes A$ must be direct; in other words, every element of T is A-invariant. Now the fixed points of A are as follows:

For $x \mapsto -x$, the points of order 2.
For $F_i = \mathbb{C}/(\mathbb{Z} \oplus \mathbb{Z}i)$ and $A = \langle x \mapsto ix \rangle$, the points $0, \frac{1+i}{2}$.
For $F_\rho = \mathbb{C}/(\mathbb{Z} \oplus \mathbb{Z}\rho)$ and $A = \langle x \mapsto \rho x \rangle$, the points $0, \pm(\frac{1-\rho}{3})$.
For F_ρ and $A = \langle x \mapsto -\rho x \rangle$, the point 0.

Moreover, since $T \times A$ is a group of translations of E, it can be generated by 2 elements; hence $T \times A \not\cong F_2 \times (\mathbb{Z}/2\mathbb{Z})$, where F_2 is the group of points of F of order 2. So we have the following results:

List VI.20 (Bagnera–de Franchis) *Let E, F be elliptic curves and G a group of translations of E acting on F. Then every bielliptic surface is of one of the following types:*

(1) $(E \times F)/G$, $G = \mathbb{Z}/2\mathbb{Z}$ acting on F by symmetry.
(2) $(E \times F)/G$, $G = \mathbb{Z}/2\mathbb{Z} \oplus \mathbb{Z}/2\mathbb{Z}$ acting on F by $x \mapsto -x$, $x \mapsto x + \epsilon$ $(\epsilon \in F_2)$.
(3) $(E \times F_i)/G$, $G = \mathbb{Z}/4\mathbb{Z}$ acting on F by $x \mapsto ix$.
(4) $(E \times F_i)/G$, $G = \mathbb{Z}/4\mathbb{Z} \oplus \mathbb{Z}/2\mathbb{Z}$, acting by $x \mapsto ix$, $x \mapsto x + (\frac{1+i}{2})$.
(5) $(E \times F_\rho)/G$, $G = \mathbb{Z}/3\mathbb{Z}$ acting by $x \mapsto \rho x$.
(6) $(E \times F_\rho)/G$, $G = \mathbb{Z}/3\mathbb{Z} \oplus \mathbb{Z}/3\mathbb{Z}$ acting by $x \mapsto \rho x$, $x \mapsto x + (\frac{1-\rho}{3})$.
(7) $(E \times F_\rho)/G$, $G = \mathbb{Z}/6\mathbb{Z}$ acting by $x \mapsto -\rho x$.

One has $2K \equiv 0$ in (1) and (2), $4K \equiv 0$ in (3) and (4), $3K \equiv 0$ in (5) and (6), while $6K \equiv 0$ in (7).

Historical Note VI.21

The main result of this chapter, the characterization of ruled surfaces by 'the termination of adjunction' or by the existence of a curve C with $K.C < 0$, appears in the 1901 paper [C-E]. Later on Enriques proved the more precise form given in VI.17 ([E3], 1905). We have more or less followed the plan of his proof.

Bielliptic surfaces (known classically as 'irregular hyperelliptic surfaces of rank > 1') were classified in [B-DF].

Exercises VI.22

(1) Let $S \subset \mathbb{P}^n$ be a surface whose smooth hyperplane sections are elliptic curves. Show that S is either a del Pezzo surface S_d or S'_8 (see Exercise V.21(2)) or an elliptic ruled surface. Give examples of the latter.
(Show that $q(S) \leqslant 1$; if $q = 0$, show that $K \equiv -H$ and then apply Exercise V.21(2). If $q = 1$, apply Corollary VI.18 to prove that S is ruled.)

(2) Let $S \subset \mathbb{P}^n$ be a surface of degree d lying in no hyperplane.
   (a) Show that $d \geqslant 2n - 2$ if S is not ruled, and that $K_S \equiv 0$ if equality holds.
   (b) Show that $d \geqslant n + q - 1$ if S is ruled.

   (Use the exact sequence $0 \to \mathcal{O}_S \to \mathcal{O}_S(H) \to \mathcal{O}_H(H) \to 0$ and apply Clifford's theorem: if D is a divisor on a curve and $0 \leqslant \deg D \leqslant 2g - 2$, then $h^0(D) \leqslant \frac{1}{2}\deg D + 1$.)

(3) Show that bielliptic surfaces are characterized by $p_g = 0$, $P_{12} = q = 1$ (examine the proof of VI.15).

(4) Let S be a minimal non-ruled surface with $\chi_{\text{top}}(S) = 0$. Assume that S has a surjective morphism to a curve. Show that there are smooth curves B, F, with $g(B) = 1$, and a group G acting on B and F such that $S \cong (B \times F)/G$.

VII
KODAIRA DIMENSION

In the preceding chapter we saw the importance of the pluricanonical linear systems $|nK|$ in the classification of surfaces. In this short chapter we are going to systematize this point of view by dividing algebraic surfaces into four classes, according to the 'ampleness' of their canonical divisor.

Definition VII.1 *Let V be a smooth projective variety, K a canonical divisor of V, ϕ_{nK} the rational map from V to the projective space associated with the system $|nK|$. The Kodaira dimension of V, written $\kappa(V)$, is the maximum dimension of the images $\phi_{nK}(V)$, for $n \geqslant 1$.*

Recall that the image of a rational map is well defined (II.4); if $|nK| = \emptyset$ then $\phi_{nK}(V) = \emptyset$, and we say $\dim(\emptyset) = -\infty$. For a curve it is easy to give the Kodaira dimension explicitly:

Example VII.2 Let C be a smooth curve of genus g. Then

$$\begin{aligned} \kappa(C) = -\infty &\Leftrightarrow g = 0 \\ \kappa(C) = 0 &\Leftrightarrow g = 1 \\ \kappa(C) = 1 &\Leftrightarrow g \geqslant 2 . \end{aligned}$$

We now interpret the definition for a surface.

Example VII.3 Let S be a surface. Then

$\kappa(S) = -\infty \Leftrightarrow P_n = 0$ for all $n \geqslant 1 \Leftrightarrow S$ is ruled (Enriques' theorem).

$\kappa(S) = 0 \Leftrightarrow P_n = 0$ or 1, and there exists N such that $P_N = 1$.

$\kappa(S) = 1 \Leftrightarrow$ there exists N such that $P_N \geqslant 2$, and for all n, $\phi_{nK}(S)$ is at most a curve.

$\kappa(S) = 2 \Leftrightarrow$ for some N, $\phi_{NK}(S)$ is a surface.

In the coming chapters we are going to classify the surfaces with $\kappa = 0, 1, 2$. For the moment we will content ourselves with a few examples.

Proposition VII.4 *Let C, D be two smooth curves, $S = C \times D$. Then*

If C or D is rational, then S is ruled (i.e. $\kappa(S) = -\infty$).
If C and D are elliptic, $\kappa(S) = 0$.
If C is elliptic and $g(D) \geqslant 2$, $\kappa(S) = 1$.
If C and D are of genus $\geqslant 2$, then $\kappa(S) = 2$.

Proof If p and q are the two projections of S to C and D, we have

$$\Omega_S^2 \cong p^*\omega_C \otimes q^*\omega_D \quad \text{and} \quad H^0(S, \mathcal{O}_S(nK)) \cong H^0(C, \omega_C^{\otimes n}) \otimes H^0(D, \omega_D^{\otimes n}),$$

so that the rational map $\phi_{nK} : S \dashrightarrow \mathbb{P}^N$ factorizes as

$$\phi_{nK} : C \times D \xrightarrow{(\phi_{nK_C}, \phi_{nK_D})} \mathbb{P}^{N'} \times \mathbb{P}^{N''} \overset{s}{\hookrightarrow} \mathbb{P}^N,$$

where s is the Segre embedding (defined by $((X_i), (Y_j)) \mapsto (X_i Y_j)$). The proposition then follows easily from VII.2.

Proposition VII.5 *Let $S_{d_1,\dots,d_r}$ denote a surface in $\mathbb{P}^{r+2}$ which is the complete intersection of r hypersurfaces of degrees $d_1, \dots, d_r$. Then:*

The surfaces S_2, S_3, $S_{2,2}$ are rational (so have $\kappa = -\infty$).
The surfaces S_4, $S_{2,3}$, $S_{2,2,2}$ have $K \equiv 0$ and $\kappa = 0$.
All other surfaces $S_{d_1,\dots,d_r}$ have $\kappa = 2$.

Proof We have $K = kH$, where H is a hyperplane section of the surface (with the given embedding) and $k = (\sum d_i) - r - 3$ (Lemma IV.11). For $k < 0$ we get the surfaces S_2, S_3, and $S_{2,2}$, which are rational (cf. IV.13 and IV.16). The surfaces with $k = 0$ (so $K \equiv 0$) are S_4, $S_{2,3}$, $S_{2,2,2}$; for all the other complete intersections the canonical divisor is a positive multiple of the hyperplane section, so $\kappa = \dim \phi_K(S) = 2$.

Finally, note that the surfaces for which $nK \equiv 0$ for some n (for example the Abelian surfaces, and the bielliptic surfaces (VI.19)) clearly have $\kappa = 0$.

Historical Note VII.6

The importance of the plurigenera in the classification of surfaces is made clear by the theorems of Castelnuovo and Enriques, characterizing rational and ruled surfaces. The division of surfaces according to the four values of κ appears in the article [E4] under a different formulation, in terms of the plurigenus P_{12} (cf. Exercise VIII.22(3)):

$$\begin{aligned} \kappa = -\infty \quad &\Leftrightarrow \quad P_{12} = 0 \\ \kappa = 0 \quad &\Leftrightarrow \quad P_{12} = 1 \\ \kappa = 1 \quad &\Leftrightarrow \quad P_{12} \geqslant 2,\ K^2 = 0 \\ \kappa = 2 \quad &\Leftrightarrow \quad P_{12} \geqslant 2,\ K^2 > 0\ . \end{aligned}$$

This article also contains the more precise classification to be described in the next chapters, which is now called Enriques' classification. The invariant κ was introduced much later; it appears for the first time – to my knowledge – in Shafarevich's seminar [Sh2].

Exercises VII.7

(1) Let V be a smooth projective variety. Consider the algebra $\bigoplus_{n\geqslant 0} H^0(V, \mathcal{O}_V(nK))$ and write d for its Krull dimension (i.e. the transcendence degree over $\mathbb{C}$ of its field of fractions). Show that:

$$\kappa(V) = d - 1 \quad \text{if} \quad \kappa(V) \geqslant 0\ .$$

(2) If V, W are two smooth projective varieties, show that $\kappa(V \times W) = \kappa(V) + \kappa(W)$.

(3) Let $f : V \to W$ be a surjective morphism of smooth projective varieties. Then $\kappa(W) \leqslant \kappa(V)$, with equality if f is étale.

VIII

SURFACES WITH $\kappa = 0$

For the next three chapters we will be studying those surfaces with $\kappa = 0, 1$ and 2. Since these surfaces have a unique minimal model, we can restrict ourselves to considering minimal surfaces. We will constantly use the fact that $K.D \geqslant 0$ for every effective divisor D on a minimal surface with $\kappa \geqslant 0$ (this is the easy part of Corollary VI.18: if C is a curve with $C.K < 0$, we have $C^2 \geqslant 0$, and hence $|nK| = \emptyset$ for all $n \geqslant 1$ by the useful remark III.5).

To say that $\kappa = 0$ means that $P_n = 0$ or 1 for all n, and there exists $m \geqslant 1$ with $P_m = 1$.

Lemma VIII.1 *Let S be a minimal surface with $\kappa = 0$. Then:*

(a) $K^2 = 0$;

(b) $\chi(\mathcal{O}_S) \geqslant 0$;

(c) *let n and m be two positive integers such that $P_n = P_m = 1$; then if d is the g.c.d. of n and m, $P_d = 1$.*

Proof (a) For some $m \geqslant 1$, $|mK|$ contains an effective divisor D; since $D.K \geqslant 0$, $K^2 \geqslant 0$. Suppose $K^2 > 0$; the Riemann–Roch theorem gives

$$h^0(nK) + h^0((1-n)K) \geqslant \chi(\mathcal{O}_S) + \binom{n}{2} K^2 .$$

For $n \geqslant 2$, $|(1-n)K| = \emptyset$, since otherwise $E \in |(1-n)K|$ would give $E.K \geqslant 0$, and hence $K^2 \leqslant 0$; we thus have that $h^0(nK) = P_n$ tends to infinity with n, contradicting $\kappa = 0$.

(b) Using $K^2 = 0$, Noether's formula becomes

$$12\chi(\mathcal{O}_S) = \chi_{\text{top}}(S) = 2 - 4q + b_2 ,$$

or

$$8\chi(\mathcal{O}_S) = -2 - 4p_g + b_2 \geqslant -2 - 4p_g \geqslant -6 ,$$

(since $p_g \leqslant 1$), hence $\chi(\mathcal{O}_S) \geqslant 0$.

(c) Let $D \in |nK|$ and $E \in |mK|$; write $n = n'd$, $m = m'd$ with $(n', m') = 1$. The two divisors $m'D$ and $n'E$ belong to the same 0-dimensional linear system $|\frac{nm}{d}K|$, and are therefore equal; from this one sees that there is an effective divisior Δ such that

$$D = n'\Delta , \quad E = m'\Delta .$$

Write $\epsilon \in \operatorname{Pic} S$ for the class of $\Delta - dK$; since $D \in |nK|$ and $E \in |mK|$, we have $n'\epsilon = m'\epsilon = 0$, and since $(n', m') = 1$, $\epsilon = 0$. Hence $\Delta \in |dK|$ and $P_d = 1$.

Theorem (and Definitions) VIII.2 *Let S be a minimal surface with $\kappa = 0$. Then S belongs to one of the 4 following cases:*

(1) $p_g = 0$, $q = 0$; *then $2K \equiv 0$, and we say that S is an 'Enriques surface'.*

(2) $p_g = 0$, $q = 1$; *then S is a bielliptic surface* (VI.19).

(3) $p_g = 1$, $q = 0$; *then $K \equiv 0$, and we say that S is a 'K3 surface'.*

(4) $p_g = 1$, $q = 2$; *then S is an Abelian surface.*

Proof We have $p_g \leqslant 1$; suppose first that $p_g = 0$.

If $q = 0$, $P_2 \geqslant 1$ by Castelnuovo's criterion V.1; Riemann–Roch, with $K^2 = 0$, gives

$$h^0(-2K) + h^0(3K) \geqslant 1 .$$

Since $p_g = 0$, VIII.1(c) gives $P_3 = 0$, and hence $h^0(-2K) \geqslant 1$; it follows that $2K \equiv 0$.

The minimal surfaces with $p_g = 0$, $q \geqslant 1$ have been classified in Chapter VI; from Proposition VI.15 one sees that the only surfaces among them with $\kappa = 0$ are the bielliptic surfaces.

Now suppose that $p_g = 1$; it follows from VIII.1(b) that $q = 0, 1$ or 2.

If $q = 0$, Riemann–Roch gives $h^0(-K) + h^0(2K) \geqslant 2$, and hence $h^0(-K) = 1$ and $K \equiv 0$.

Suppose that $q = 1$. By I.10 this implies that there is $\epsilon \in \operatorname{Pic} S$ with $\epsilon \not\equiv 0$, but $2\epsilon \equiv 0$; in particular, $\epsilon.D = 0$ for every divisor D on S, and $h^0(\epsilon) = h^0(-\epsilon) = 0$. Now Riemann–Roch becomes

$$h^0(\epsilon) + h^0(K - \epsilon) \geqslant 1 ,$$

and hence $h^0(K - \epsilon) \geqslant 1$. Let $D \in |K - \epsilon|$ and $E \in |K|$; then $2D$, $2E \in$

$|2K|$, giving $2D = 2E$, since $P_2 = 1$. But then $D = E$, which contradicts $\epsilon \not\equiv 0$, and so there does not exist any surface with $p_g = P_2 = q = 1$.

It remains to prove (4). Firstly, we note the following important proposition.

Proposition VIII.3 *Let S be a surface, C_i irreducible curves on S, and $m_i > 0$ integers. Set $F = \sum m_i C_i$, and suppose that for each i, $F.C_i \leqslant 0$. Let $D = \sum r_i C_i$ with $r_i \in \mathbb{Z}$ (or even $r_i \in \mathbb{Q}$), and $D \neq 0$. Then*

(a) $D^2 \leqslant 0$;

(b) *if F is connected and $D^2 = 0$, one has $D = rF$ for some $r \in \mathbb{Q}$, and $F.C_i = 0$ for all i.*

This means that the intersection matrix $(C_i.C_j)$ is negative semi-definite, and that its kernel has dimension at most 1 if F is connected.

Proof Set $G_i = m_i C_i$, and $s_i = r_i/m_i$ so that $F = \sum G_i$ and $D = \sum s_i G_i$. Then

$$D^2 = \sum_i s_i^2 G_i^2 + 2\sum_{i<j} s_i s_j G_i G_j \ .$$

Write $G_i^2 = G_i.(F - \sum_{j\neq i} G_j)$:

$$\begin{aligned} D^2 &= \sum_i s_i^2 G_i.F - \sum_{i<j}(s_i^2 + s_j^2 - 2s_i s_j)G_i.G_j \\ &= \sum_i s_i^2 G_i.F - \sum_{i<j}(s_i - s_j)^2 G_i.G_j \ . \end{aligned}$$

(a) is now clear. If $D^2 = 0$, we must have $s_i = s_j$ every time that $C_i \cap C_j \neq \emptyset$. Suppose F is connected; then any two components C_i and C_j can always be joined by a chain of C_k; it follows that all the s_i are equal (hence non-zero). Moreover, we must also have $s_i^2 G_i.F = 0$ for each i, and hence $C_i.F = 0$.

We have in mind two applications of the proposition.

Corollary VIII.4 *Let S be a surface, B a smooth curve, and $p: S \to B$ a surjective morphism with connected fibres. Suppose that $p^*b = F_b = \sum_i m_i F_i$ for some $b \in B$; then if $D = \sum_i r_i F_i$, with $r_i \in \mathbb{Z}$, we have $D^2 \leqslant 0$, with equality if and only if $D = rF_b$ for some $r \in \mathbb{Q}$.*

Proof This can be read off directly from the proposition (with $F_i = C_i$, $F_b = F$).

Corollary VIII.5 *Let S be a surface, $g : S \to S'$ a surjective morphism to a (possibly singular) projective surface S', and let C_i be irreducible curves of S such that $g(C_i) = p \in S'$. Then for any $D = \sum_i r_i C_i$, with $r_i \in \mathbb{Z}$, one has $D^2 < 0$.*

Proof It is enough to prove the corollary when $\cup C_i$ is connected, applying it subsequently to each component. Replacing g by its Stein factorization (V.17), we can assume that g has connected fibres. We then have $g^{-1}(p) = \bigcup_i C'_i$, where the C'_i are irreducible curves, including the C_i. Let H be a hyperplane section of S' passing through p. Write

$$g^*H = \tilde{H} + F \ , \quad F = \sum_i m_i C'_i \ ,$$

with $\tilde{H}$ not containing any C'_i. Then $g^*H.C'_i = 0$ for each i, since we can move H away from p; hence $F.C'_i = (g^*H - \tilde{H}).C'_i \leqslant 0$ for all i, with strict inequality for at least one i, since H meets the fibre $g^{-1}(p)$. The proposition now implies the result.

VIII.6 Proof of Theorem VIII.2(4) Let S be a minimal surface with $\kappa = 0$, $p_g = 1$, $q = 2$; we denote by K the effective canonical divisor $K \in |K|$.

(α) *Structure of K*

Suppose $K \neq 0$. Then $K = \sum_i n_i C_i$ where the C_i are irreducible curves, and $n_i > 0$; because $K^2 = 0$, and $K.C_i \geqslant 0$ for each i, we must have $K.C_i = 0$. Replacing K by $\sum n_i C_i$, we get $n_i C_i^2 + \sum_{j \neq i} n_j C_i.C_j = 0$, and hence
either $C_i^2 < 0$, so $C_i^2 = -2$ and $C_i \cong \mathbb{P}^1$;
or C_i^2 and $C_i.C_j = 0$ for all $j \neq i$. Then C_i is a smooth elliptic curve or a rational curve with a double point, and is a connected component of $\cup C_i$.

Write $K = \sum D_\alpha$, where D_α are connected effective divisors and $D_\alpha.D_\beta = 0$ for $\alpha \neq \beta$ (the D_α will be called the connected components of K). From what we have seen, $D_\alpha^2 = 0$, and each D_α is
either a smooth elliptic curve, a rational curve with a double point, or a multiple of one of these;
or a union of smooth rational curves.
We will now consider the Albanese morphism $\alpha : S \to \text{Alb}(S)$ and prove that it is an étale cover. In $(\beta) - (\epsilon)$ below we consider the 4 different possible cases.

(β) *$\alpha(S)$ is a curve and $K \neq 0$*
Then (V.15) $\alpha(S) = B$ is a smooth curve of genus 2, and the fibration $p : S \to B$ has connected fibres. Now no curve of genus 0 or 1 maps surjectively to B, so that every connected component D of K is contained in a fibre F_b of p. Since $D^2 = 0$, this together with Corollary VIII.4 implies that $D = \frac{m}{q} F_b$, with m and q positive integers. But then

$$rq\, D = rm\, F_b = p^*(rm[b])\ ;$$

hence $h^0(nD)$, and with it also $h^0(nK)$, tends to infinity with n, which contradicts $\kappa = 0$.

(γ) *$\alpha(S)$ is a curve and $K = 0$*
We keep the notation of (β). Consider an étale cover $B' \to B$ of degree $\geqslant 2$, and set $S' = S \times_B B'$. Then S' is connected (since the fibres of p are connected); moreover, $K_{S'} \equiv \pi^* K_S \equiv 0$ and $\chi(\mathcal{O}_{S'}) = 0$ (by Lemma VI.3), which implies $q(S') = 2$; but $q(S') \geqslant g(B')$, and by the Hurwitz formula $g(B') \geqslant 3$, which is a contradiction.

(δ) *α is surjective and $K \neq 0$*
Let D be a connected component of K. Since $D^2 = 0$, Corollary VIII.5 shows that α does not contract D to a point, and so D cannot be a union of rational curves. By (α), it follows that $D = nE$, where E is a smooth elliptic curve, and $E' = \alpha(E)$ is a smooth elliptic curve in Alb(S). After a suitable translation if necessary, we can suppose that E' is a sub-Abelian variety in Alb(S) (see V.12). Consider the quotient curve $F = \mathrm{Alb}(S)/E'$ and the composite morphism $f : S \to F$, and let $S \xrightarrow{g} B \to F$ be its Stein factorization (V.17). The curve E being contained in a fibre F_b of g, Corollary VIII.4 gives $E = \frac{1}{q} F_b$ (for q a positive integer), and as before $h^0(mD) \to \infty$ with m; hence also $h^0(mK) \to \infty$ with m, a contradiction.

There then only remains one possibility.

(ϵ) *α is surjective and $K = 0$*
Let (η_1, η_2) be a basis of $H^0(A, \Omega^1_A)$; set $\omega_1 = \alpha^* \eta_1$ and $\omega_2 = \alpha^* \eta_2$. Since η_1, η_2 form a basis of Ω^1_A everywhere on A, α is étale at $x \in S$ if and only if the 2-form $\omega_1 \wedge \omega_2$ does not vanish at x. Now since α is generically étale, $\omega_1 \wedge \omega_2$ is not identically zero (this is where we use characteristic zero: the proof in characteristic p is much more delicate, see [B-M]); since $K \equiv 0$ it is then nowhere zero. Hence α is an étale cover; since every étale cover

of an Abelian variety is again an Abelian variety, S is an Abelian surface. This completes the proof of Theorem VIII.2.

Corollary VIII.7 *A minimal surface with $\kappa = 0$ satisfies $4K \equiv 0$ or $6K \equiv 0$.*

Indeed, for $K3$ and Abelian surfaces $K \equiv 0$, while Enriques surfaces have $2K \equiv 0$ and the bielliptic surfaces $4K \equiv 0$ or $6K \equiv 0$ (see VI.20).

The further study of Abelian surfaces relates more to the general theory of Abelian varieties than to surface theory, and we will say no more about it, limiting ourselves to a reference to [M3]. Bielliptic surfaces have been treated in VI.20; in the remainder of this chapter we will give some examples of $K3$ and Enriques surfaces.

K3 surfaces

By definition these are the surfaces with $K \equiv 0$, $q = 0$. Clearly, $K \equiv 0$ implies that they are minimal; Noether's formula gives $\chi_{\text{top}} = 24$, whence $b_2 = 22$.

Example VIII.8 Complete intersections

In VII.5 we have seen that the only complete intersections with $K \equiv 0$ are quartic surfaces $S_4 \subset \mathbb{P}^3$, $S_{2,3} \subset \mathbb{P}^4$ and $S_{2,2,2} \subset \mathbb{P}^5$. These are $K3$s by the following lemma:

Lemma VIII.9 *Let $V \subset \mathbb{P}^n$ be a d-dimensional complete intersection; then $H^i(V, \mathcal{O}_V) = 0$ for $0 < i < d$.*

Proof We prove that in fact, more generally, $H^i(V, \mathcal{O}_V(k)) = 0$ for $0 < i < d$ and any $k \in \mathbb{Z}$. By induction on the number of equations defining V, it is enough to prove that if this statement holds for V, it holds also for the section W of V by a hypersurface of degree r. This then follows from the cohomology long exact sequence of

$$0 \to \mathcal{O}_V(k-r) \to \mathcal{O}_V(k) \to \mathcal{O}_W(k) \to 0\ .$$

Example VIII.10 Kummer surfaces

Let A be an Abelian surface; if we choose an origin, A acquires a group structure. Let τ be the involution of A given by $a \mapsto -a$. The fixed points of τ are the points of order 2 of the group A, which is isomorphic as a group to $(\mathbb{R}/\mathbb{Z})^4$; there are thus 16 points of order 2, $p_1, \ldots, p_{16}$. Let $\epsilon : \hat{A} \to A$ be the blow-up of these 16 points, and $E_i = \epsilon^{-1}(p_i)$

the exceptional curves; the involution τ extends to an involution σ of $\hat{A}$; denote by X the quotient $X = \hat{A}/\{1, \sigma\}$, and let $\pi : \hat{A} \to X$ be the projection.

Proposition VIII.11 *$X = \hat{A}/\{1, \sigma\}$ is a $K3$ surface, the 'Kummer surface' of A.*

Proof We show first that X is smooth; π is étale outside the E_i, so that it is enough to check this at points $\pi(q)$, where $q \in E_i$. Writing A in the form V/Γ as in V.11, we get local coordinates (x, y) on A in a neighbourhood of p_i, such that $\tau^* x = -x$, $\tau^* y = -y$. Set $x' = \epsilon^* x$, $y' = \epsilon^* y$; we can suppose that x' and $t = y'/x'$ are local coordinates on $\hat{A}$ near q. Now $\sigma^* x' = -x'$ and $\sigma^* t = (-y')/(-x') = t$, so that t and $u = x'^2$ form a system of local coordinates on X near $\pi(q)$: thus X is smooth.

Now we compute the canonical divisor of X. A has a holomorphic 2-form ω which is nowhere zero; if $A = V/\Gamma$, and x, y are coordinates on V, then up to a scalar factor, $\omega = dx \wedge dy$, whence $\tau^* \omega = \omega$. The 2-form $\epsilon^* \omega$ is thus invariant under σ; using Lemma VI.11, it follows that $\epsilon^* \omega = \pi^* \alpha$, where α is a meromorphic 2-form on X. Clearly, the divisor of α is concentrated on the E_i. Let $q \in E_i$; in the above notation, in a neighbourhood of q we have

$$\epsilon^* \omega = dx' \wedge dy' = dx' \wedge d(tx') = x' dx' \wedge dt = \frac{1}{2} du \wedge dt \,.$$

Thus α is holomorphic and non-zero at q, proving that $\operatorname{div}(\alpha) = 0$, whence $K_X \equiv 0$.

Finally, if X has a non-zero holomorphic 1-form, we could deduce that $\hat{A}$ has a 1-form invariant under σ; since $\epsilon^* : H^0(A, \Omega^1_A) \to H^0(\hat{A}, \Omega^1_{\hat{A}})$ is an isomorphism (III.20), this would imply that A has a 1-form invariant under τ, which is absurd, since $H^0(A, \Omega^1_A)$ has $\{dx, dy\}$ as a basis. Thus $q(X) = 0$ and X is a $K3$ surface.

Remarks VIII.12

(1) One can also consider the singular quotient $X' = A/\tau$; this has 16 ordinary double points r_i, the images of the p_i. From ϵ we can construct a morphism $\epsilon : X \to X'$ which is an isomorphism outside the E_i, and contracts E_i to r_i.

(2) Suppose that A is the Jacobian of a curve C of genus 2; C can be embedded in A in such a way that $\tau(C) = C$: if $r \in C$ is a Weierstrass point, then $2r \equiv K_C$, and we can take the embedding

$x \mapsto [x] - [r]$. It is not difficult to show (see Exercise 4) that the system $|2C|$ on A defines a morphism $k : A \to \mathbb{P}^3$ which factors through the quotient by τ, and defines an embedding $X' \hookrightarrow \mathbb{P}^3$. Since $(2C)^2 = 8$, the image $X' \subset \mathbb{P}^3$ is a surface of degree 4 with 16 ordinary double points. It is this surface which is known classically as the Kummer surface (see Exercises 4–9).

(3) The E_i for $1 \leqslant i \leqslant 16$ form an orthogonal system in $\operatorname{Pic} X = NS(X)$; $\operatorname{Pic} X$ also contains the class of a hyperplane section, so that $\operatorname{Pic} X$ has rank $\geqslant 17$ as $\mathbb{Z}$-module. It can be shown that the rank of $\operatorname{Pic} X$ is $\leqslant 20$ for any $K3$.

We are now going to study linear systems on a $K3$ surface and the morphisms to projective space which they define.

Proposition VIII.13 *Let S be a $K3$ surface, $C \subset S$ a smooth curve of genus g. Then*

(i) *$C^2 = 2g - 2$ and $h^0(C) = g + 1$.*

(ii) *If $g \geqslant 1$ the system $|C|$ is without base points, so defines a morphism $\phi : S \to \mathbb{P}^g$; the restriction of ϕ to C is the canonical morphism $C \to \mathbb{P}^{g-1}$ defined by $|\omega_C|$.*

(iii) *If $g = 2$, $\phi : S \to \mathbb{P}^2$ is a morphism of degree 2, whose branch locus is a sextic of $\mathbb{P}^2$.*

(iv) *Suppose $g \geqslant 3$; then*

either

ϕ is a birational morphism, and a generic curve of $|C|$ is non-hyperelliptic;

or

ϕ is a 2-to-1 morphism to a rational surface (possibly singular) of degree $g - 1$ in $\mathbb{P}^g$. A generic curve of $|C|$ is then hyperelliptic.

(v) *If $g \geqslant 3$ (resp. $g = 2$), the morphism ϕ' defined by $|2C|$ (resp. $|3C|$) is birational.*

Proof

(i) The genus formula gives $C^2 = 2g - 2$; using Riemann–Roch, the second formula is equivalent to $h^1(C) = 0$, and in turn to $h^1(-C) = 0$, since $K \equiv 0$. This follows at once from the exact sequence

$$0 \to \mathcal{O}_S(-C) \to \mathcal{O}_S \to \mathcal{O}_C \to 0 .$$

(ii) Since $K \equiv 0$ and $\mathcal{O}_S(K+C)_{|C} \cong \omega_C$ by I.16(ii), there is an exact sequence

$$0 \to \mathcal{O}_S \to \mathcal{O}_S(C) \to \omega_C \to 0 ,$$

showing that $|C|$ cuts out on C the complete linear system $|\omega_C|$. This has no base points on C, and hence $|C|$ has no base points. The final assertion in (ii) is clear.

(iii) If $g = 2$, we have $C^2 = 2$, so ϕ has degree 2. Let $\Delta \subset \mathbb{P}^2$ denote the branch curve of ϕ, $\deg \Delta = n$. Now $C = \phi^{-1}(\ell)$, where $\ell = \phi(C)$ is a line in $\mathbb{P}^2$; C is then a double cover of ℓ branched in the n points of $\ell \cap \Delta$. Since $g(C) = 2$, $n = 6$.

(iv) If C is non-hyperelliptic, the restriction to C of ϕ is an embedding; since $\phi^{-1}(\phi(C)) = C$, this implies that ϕ is birational.

If ϕ is not birational, every smooth curve in $|C|$ is then hyperelliptic; it follows that for a generic $x \in S$, $\phi^{-1}(\phi(x))$ consists of 2 points, and hence ϕ has degree 2. Since $C^2 = 2g-2$, the image of ϕ is a (possibly singular) surface Σ of degree $g-1$ in $\mathbb{P}^g$, whose hyperplane sections are the rational curves $\phi(C)$; it follows at once that Σ is rational (see also Chapter IV, Exercise 4).

(v) The restriction to C of ϕ' is the 2-canonical (resp. 3-canonical) morphism, which is an embedding; as before this implies that ϕ' is birational.

The reader familiar with duality and the properties of linear systems on singular curves will note that the proof works also if C is merely assumed to be irreducible.

Example VIII.14 We restrict ourselves to the case when ϕ is birational (see Exercise 10 for the hyperelliptic case).

If $g = 3$, $\phi(S) \subset \mathbb{P}^3$ is a quartic.

If $g = 4$, $\phi(S)$ is a surface in $\mathbb{P}^4$ of degree 6; $h^0(\mathcal{O}_S(2C)) = 14$ by VIII.13(i), and $h^0(\mathbb{P}^4, \mathcal{O}(2)) = 15$, so that $\phi(S)$ is contained in a quadric. Since also $h^0(S, \mathcal{O}(3C)) = 29$ and $h^0(\mathbb{P}^4, \mathcal{O}(3)) = 35$, it follows that $\phi(S)$ is a complete intersection $S_{2,3}$.

If $g = 5$, then $\phi(S)$ is a surface in $\mathbb{P}^5$ of degree 8, contained in 3 linearly independent quadrics. It can be seen (Exercise 11) that in the generic case, $\phi(S)$ is a complete intersection of 3 quadrics.

We have not given precise statements on the properties of ϕ; it can be shown that ϕ is an isomorphism outside certain divisors, which it contracts to singularities of $\phi(S)$ of a very special type, called rational double points.

We have thus discussed the structure of $K3$s of degree $2g-2$ in $\mathbb{P}^g$, for $g = 3,4,5$; such surfaces exist for arbitrary $g \geqslant 3$, as we now prove.

Proposition VIII.15 *For every $g \geqslant 3$, there exist $K3$ surfaces $S = S_{2g-2}$ embedded in $\mathbb{P}^g$.*

Proof In view of VIII.13(i), it is enough to construct $K3$ surfaces S containing a very ample divisor D with $D^2 = 2g-2$ (recall that a divisor is very ample if the corresponding linear system $|D|$ defines an embedding of S into $|D|^\vee$). Note first that if H is a hyperplane section of S and $|E|$ a linear system without base points, $H+E$ is a very ample divisor: if x, y are distinct points of S, there is a divisor $E_0 \in |E|$ not containing either x or y, and $H_0 \in |H|$ containing x but not y. Thus $|H+E|$ separates points, and the same argument shows that it separates tangent directions.

We now distinguish 3 cases:

• $g = 3k$, with $k \geqslant 1$. Let $S \subset \mathbb{P}^3$ be a quartic containing a line ℓ, and let $|E|$ be the pencil of elliptic curves $|H - \ell|$; then $D_k = H + (k-1)E$ is very ample, and we have

$$D_k^2 = 4 + 6(k-1) = 2g - 2 \ .$$

• $g = 3k+1$, with $k \geqslant 1$. Let $Q \subset \mathbb{P}^4$ be a quadric with an ordinary double point (that is Q is a cone over a non-singular quadric of $\mathbb{P}^3$); let $V \subset \mathbb{P}^4$ be a cubic such that $S = Q \cap V$ is a smooth surface. Consider one of the two pencils of planes on Q: it cuts out on V a pencil of elliptic curves $|E|$. Then $D_k = H + (k-1)E$ is very ample on S, and we have

$$D_k^2 = 6 + 6(k-1) = 2g - 2 \ .$$

• $g = 3k+2$, with $k \geqslant 1$. Let $S \subset \mathbb{P}^3$ be a smooth quartic containing a line ℓ and a twisted cubic t disjoint from ℓ (for example, the equation $Y^3(X+T) + Z^3(T-X) - XT(X^2+T^2) = 0$ defines a smooth quartic containing the line $X = T = 0$, and also the twisted cubic given parametrically by $X = U^3$, $Y = U^2V$, $Z = UV^2$, $T = V^3$). Set $E = H - \ell$ and $H' = 2H - t$. Since $H'^2 = 2$, $|H'|$ defines a double cover $\pi : S \to \mathbb{P}^2$ by VIII.13(iii); for $u \in \mathbb{P}^2$, the points of $\pi^{-1}(u)$ are the intersection of S with the residual intersection of 2 quadrics of $\mathbb{P}^3$ containing t, that is, the intersection of S with a bisecant of t: the pairs of points identified by π are thus the pairs (x,y) such that the line $\langle x,y \rangle$ is a bisecant. To show that $H' + E$ is very ample, it is enough to show that there does not exist any divisor $E_0 \in |E|$ containing such a pair (x, y). But if this happened, $\langle x, y \rangle$ would cut S again in two points of t, which would then also be points of E_0; we would then have $x, y \in E_0$, for which $\langle x, y \rangle$ would be contained in E_0, and hence also in S. However, direct computation shows that this does not occur in the given example.

Thus $D_k = H' + kE$ is very ample for $k \geqslant 1$, and we have

$$D_k^2 = 2 + 6k = 2g - 2 ,$$

which completes the proof of the proposition.

Remark VIII.16

We have seen that for $g = 3, 4, 5$, $K3$ surfaces of degree $2g - 2$ in $\mathbb{P}^g$ form an irreducible family, that is, can be parametrized by an irreducible variety T_g: for example, T_3 is an open subset of the projective space $|\mathcal{O}_{\mathbb{P}^3}(4)|$.

It is instructive to 'count the moduli' of $K3$s in $\mathbb{P}^g$, in the following entirely heuristic manner: the projective group $PGL(g+1)$ acts on T_g; it can be seen that the stabilizer of a point is finite (see Exercise V.21(3)). It is therefore natural to take the number of moduli to be $m_g = \dim T_g - \dim PGL(g+1)$. In the case $g = 2$, the surfaces one gets are double covers of $\mathbb{P}^2$ branched in a sextic; since plane sextics depend on 27 parameters, we get $m_2 = 19$. For $g = 3$, we get $m_3 = \binom{4+3}{3} - 1 - 15 = 19$; similarly, one finds $m_4 = m_5 = 19$.

This heuristic calculation is justified by deformation theory. We limit ourselves to a brief description, referring for example to [Sh2, Chapter 9] for details and the proofs. Every compact analytic manifold X has a 'local moduli space' parametrizing small deformations of X. If $H^2(X, T_X) = 0$, this space is smooth and of dimension $h^1(T_X)$.

In the case of a $K3$ surface, $H^2(X, T_X) = H^0(X, \Omega_X^1)^\vee$ by duality, and $h^1(T_X) = h^1(\Omega_X^1) = b_2 - 2p_g = 20$ by Hodge theory. The local moduli space of X is hence 20-dimensional. The discrepancy between this and the geometric calculation comes from the fact that a surface obtained as a deformation of an algebraic $K3$ surface may no longer be algebraic. The local moduli space of X thus consists of an open subset U of $\mathbb{C}^{20}$ containing a countable infinity of 19-dimensional subvarieties which correspond to algebraic $K3$s; the union of these is dense in U.

Enriques surfaces

Recall first that if X is a variety and $L \in \operatorname{Pic} X$ is an invertible sheaf of order 2 (that is, there is an isomorphism $\alpha : L^{\otimes 2} \xrightarrow{\sim} \mathcal{O}_X$), L corresponds to an étale double cover $\pi : \tilde{X} \to X$, characterized by the fact that $\pi^* L \cong \mathcal{O}_{\tilde{X}}$. Viewing L as a line bundle, one can take

$$\tilde{X} = \{u \in L \mid \alpha(u^{\otimes 2}) = 1\} ,$$

the cover π being defined by the projection of L to X. Sending a point u of $\tilde{X} \subset L$ to the point (u, u) in $\tilde{X} \times_X L = \pi^* L$ defines a nowhere-vanishing section of $\pi^* L$, which is hence trivial.

Proposition VIII.17 *Let S be an Enriques surface, and $\pi : \tilde{S} \to S$ the étale double cover corresponding to the invertible sheaf $\mathcal{O}_S(K)$ of order two. Then $\tilde{S}$ is a $K3$ surface.*

Conversely, the quotient of a $K3$ surface by a fixed-point free involution is an Enriques surface.

Proof From the above, $K_{\tilde{S}} \equiv \pi^* K_S \equiv 0$; since $\chi(\mathcal{O}_{\tilde{S}}) = 2\chi(\mathcal{O}_S) = 2$ by VI.3, we get $q(\tilde{S}) = 0$, so that $\tilde{S}$ is a $K3$ surface.

Conversely, let $\pi : \tilde{S} \to S$ be an étale double cover, with $\tilde{S}$ a $K3$ surface. Since $\pi^* K_S \equiv K_{\tilde{S}} \equiv 0$, one sees that $2K_S \equiv \pi_* \pi^* K_S \equiv 0$; moreover, $\chi(\mathcal{O}_S) = 1$, which is enough to prove that S is an Enriques surface, by Theorem VIII.2.

Enriques surfaces thus correspond bijectively to $K3$s together with a fixed-point free involution. Here are some examples of such involutions.

Example VIII.18 Let X be the complete intersection in $\mathbb{P}^5$ of the 3 quadrics

$$Q_i(X_0, X_1, X_2) + Q_i'(X_3, X_4, X_5) = 0$$

(for $i = 1, 2, 3$), where the Q_i and Q_i' are quadratic forms in 3 variables. We will suppose that X is a smooth surface, which holds for generic choice of Q_i, Q_i'; then by VIII.8, X is a $K3$. The involution σ of $\mathbb{P}^5$ defined by $\sigma(X_0, X_1, X_2, X_3, X_4, X_5) = (X_0, X_1, X_2, -X_3, -X_4, -X_5)$ takes X to itself; its fixed locus consists of the two planes $X_0 = X_1 = X_2 = 0$ and $X_3 = X_4 = X_5 = 0$. We will suppose that the 3 conics Q_1, Q_2 and Q_3 (resp. Q_1', Q_2', Q_3') in these planes have no points in common; this is the generic case. Under these conditions σ acts on X without fixed points, and the quotient X/σ is hence an Enriques surface.

It can be shown that the generic Enriques surface can be obtained in this way.

Example VIII.19 The Reye congruence
Let P be a linear system of quadrics of $\mathbb{P}^3$ of (projective) dimension 3. We impose the following two conditions on P, which are both satisfied if P is general enough:

(H_1) $\bigcap_{Q\in P} Q = \emptyset$;

(H_2) if ℓ is a line of $\mathbb{P}^3$ which is the vertex of a quadric $Q \in P$, then no other quadric of P contains ℓ.

Containing a line $\ell \subset \mathbb{P}^3$ will in general impose 3 linear conditions on a quadric $Q \subset \mathbb{P}^3$; ℓ will therefore in general be contained in just one quadric of P. Let S denote the variety of lines $\ell \subset \mathbb{P}^3$ contained in a pencil of quadrics of P.

Proposition VIII.20 *S is an Enriques surface.*

Proof Let X denote the subvariety of $\mathbb{P}^3 \times \mathbb{P}^3$ of pairs (x, y) such that x and y are polar with respect to all the quadrics of P. If $(x, y) \in X$, $x \neq y$ since otherwise x would belong to all Q in P. Hence the involution σ of X defined by $(x, y) \mapsto (y, x)$ does not have any fixed point on X. If $(x, y) \in X$, the quadrics of P through x and y contain the line $\langle x, y\rangle$; this line thus belongs to S. Conversely, let ℓ be a line of S; the system P induces on ℓ a pencil of (0-dimensional) quadrics, and there is exactly one pair of points (x, y) polar with respect to all the quadrics of this pencil. Hence the map $(x, y) \mapsto \langle x, y\rangle$ induces an isomorphism of X/σ with S. It is thus enough to prove that X is a $K3$ surface.

Let Q_i (for $1 \leqslant i \leqslant 4$) be 4 quadrics spanning P, and let ϕ_i be the corresponding bilinear forms. X is defined in $\mathbb{P}^3 \times \mathbb{P}^3$ by the 4 equations $\phi_i(x, y) = 0$. It is immediate by the Jacobian criterion that X is smooth and 2-dimensional at a point (x, y) if and only if the line $\langle x, y\rangle$ is not contained in the vertex of a quadric of P; this is exactly the possibility which is excluded by condition H_2. Thus $X \subset \mathbb{P}^3 \times \mathbb{P}^3$ is a surface, and is the complete intersection of 4 divisors of bidegree $(1, 1)$. It can then be shown as in the case of an ordinary complete intersection that X is a $K3$; the argument of IV.11 shows that $K_X \equiv 0$, and that of VIII.9 that $q(X) = 0$.

Enriques surfaces constructed in this way are not the most general ones (see Exercise 13).

We finally mention (without proof, although see Exercise 15) Enriques' original example: let $S \subset \mathbb{P}^3$ be a surface of degree 6 having the 6 edges of a tetrahedron as double lines, and no other singularities. Then the normalization of S is an Enriques surface.

Historical Note VIII.21

Theorem VIII.2 is contained in the paper of Enriques ([E4], 1914) we have already quoted. The fact that certain surfaces of degree $2g-2$ in $\mathbb{P}^g$ have $K \equiv 0$, $q = 0$, and in consequence have canonical curves of genus g as hyperplane sections, is observed from the very beginning of birational geometry (see for example [E6, III.6]): Enriques ([E5]) showed later that such surfaces exist for all g, and that they depend on at least 19 'moduli'; a little later Severi showed that the number of moduli is exactly 19 ([Se3]). The theory of moduli of $K3$ surfaces was considerably clarified and deepened later, in particular with the 'Torelli theorem' ([Sh-P]).

Enriques introduced the sextic with the six edges of a tetrahedron as double curves (1894, cf. [C2]), in order to give an example of an irrational surface with $p_g = q = 0$. In 1906 he returned to these surfaces ([E7]), and gave a very complete treatment of them: he proved that every surface with $P_2 = 1$, $P_3 = q = 0$ is isomorphic to a surface of the preceding type, and also that it contains a pencil of elliptic curves.

Exercises VIII.22

(1) Let S be a minimal surface with $p_g = 1$, $q = 2$ such that S has a surjective morphism to a curve of genus $\geqslant 1$. Show that:
either S is an Abelian surface;
or $P_2 > 1$.
(We know $S = (B \times F)/G$ by Chapter VI, Exercise 4, with B an elliptic curve; deal with the 2 cases B/G elliptic and B/G rational.)

(2) Let S be a minimal surface with $P_2 = 1$, $q = 2$. Show that S is an Abelian surface.
(By Theorem X.4, it follows that these conditions imply $p_g = 1$, $K^2 = \chi_{\text{top}} = 0$; reworking the proof in VIII.6, show that if S is not an Abelian surface, it has a surjective morphism to a curve of genus $\geqslant 1$; now conclude by Exercise 1.)

(3) Let S be a minimal surface. Reworking the proof of Theorem VIII.2, prove the following statements:

$p_g = 0$, $q = 0$, $P_6 = 1 \Rightarrow S$ is an Enriques surface;
$p_g = 0$, $q = 1$, $P_{12} = 1 \Rightarrow S$ is a bielliptic surface (compare Chapter VI, Exercise 3);
$p_g = 1$, $q = 0$, $P_2 = 1 \Rightarrow S$ is a $K3$ surface;
$p_g = 1$, $q = 1$ and $P_2 = 1$ is impossible;
$p_g = 1$, $q = 2$, $P_2 = 1 \Rightarrow S$ is an Abelian surface (Exercise 2).
Conclude that $\kappa = 0$ is equivalent to $P_{12} = 1$.

(4) Let C be a curve of genus 2, and JC its Jacobian. If r is a Weierstrass point of C, set $\Theta = \{[x]-[r] \mid x \in C\} \subset JC$ (compare VIII.12(2)). Show that $|2\Theta|$ defines a 2-to-1 morphism of JC to a surface of $\mathbb{P}^3$, which on passing to the quotient gives an isomorphism of JC/τ with a quartic $S \subset \mathbb{P}^3$ having 16 ordinary double points. Show that $r\Theta$ is very ample for $r \geqslant 3$.
(Recall that the map $\Theta : JC \to \operatorname{Pic}^0 JC$ of JC into the group of divisor classes algebraically equivalent to 0 on JC, defined by $\Theta(a) = \Theta_a - \Theta$, where $\Theta_a = a + \Theta$, is a group isomorphism; in particular, $\Theta_a + \Theta_{-a} \equiv 2\Theta$ for any $a \in JC$.)
In Exercises 5–9 below, a Kummer surface is taken to mean the quartic $S \subset \mathbb{P}^4$ just constructed.

(5) Show that there are 16 planes of $\mathbb{P}^3$ which touch S along a conic. Each of these conics passes through 6 of the double points of S, and each double point lies on 6 conics. The double cover of any of the 16 conics branched in its 6 double points is isomorphic to C.
(One may observe that if JC_2 denotes the group of points of order 2 on JC, the function on JC_2 which takes the value 0 on $JC_2 \cap \Theta$ and 1 on $JC_2 - \Theta$ is a quadratic form on JC_2.)

(6) Projecting a Kummer surface $S \subset \mathbb{P}^3$ from one of its double points, we obtain a (ramified) double cover of $\mathbb{P}^2$. Show that the branch locus is the union of 6 lines, which are common tangents of a conic Q; the double cover of Q branched in the 6 points of tangency is isomorphic to C. Deduce that every quartic surface in $\mathbb{P}^3$ with 16 ordinary double points is a Kummer surface.

(7) Let $p_1, \dots, p_6$ be 6 points of $\mathbb{P}^3$ such that the linear system P of quadrics through $p_1, \dots, p_6$ is 3-dimensional. Consider the following varieties:

$$\Delta = \{Q \in P \mid Q \text{ is singular}\} \subset P,$$
$$W = \{x \in \mathbb{P}^3 \mid \exists Q \in P \text{ such that } x \in \operatorname{Sing} Q\} \subset \mathbb{P}^3.$$

Show that the rational map $\pi : \mathbb{P}^3 \dashrightarrow \check{P}$ defined by the system P is of degree 2, ramified along W; its branch locus in $\check{P}$ is the dual surface of Δ (i.e. the locus of hyperplanes in P tangent to Δ). Prove that Δ is the Kummer surface associated to a Jacobian JC, and that W (classically known as the *Weddle surface* associated to $p_1, \ldots, p_6$) is birational to Δ. There is a unique twisted cubic t passing through $p_1, \ldots, p_6$; show that the double cover of t branched in the 6 p_i is isomorphic to C.

(The 16 double points of Δ correspond to the 10 quadrics of rank 2 in P, and to the 6 quadrics having $p_1, \ldots, p_6$ as vertices. Note that these 6 quadrics all belong to the (2-dimensional) system of quadrics containing t.)

(8) Representing a line of $\mathbb{P}^3$ by its 6 Plücker coordinates A, B, C, L, M, N, the variety of lines in $\mathbb{P}^3$ is identified with the quadric $G \subset \mathbb{P}^5$ having the equation $AL + BM + CN = 0$. Let Q be a further quadric of $\mathbb{P}^5$ such that $Q \cap G$ is smooth; the set of lines $Q \cap G \subset G$ is called a quadratic line complex. To each point $x \in \mathbb{P}^3$ there corresponds the cone C_x of lines of the complex through x; let $S = \{x \in \mathbb{P}^3 \mid \mathrm{rk}\ C_x \leqslant 2\}$. Show that S is a Kummer surface. Deduce that the variety of lines contained in a non-singular complete intersection of 2 quadrics in $\mathbb{P}^5$ is isomorphic to the Jacobian of a curve of genus 2.

(One may, following Klein, argue in the direction indicated; alternatively, it is perhaps easier to prove the second statement first.)

(9) Denote by $\epsilon : \hat{A} \to JC$ the blow-up of the 16 points of order 2 of JC, and let $E_1, \ldots, E_{16}$ be the exceptional curves; set $D = 4\epsilon^*\Theta - \sum_{i=1}^{16} E_i$. Show that the linear system $|D|$ defines a morphism of $\hat{A}$ into $\mathbb{P}^5$, which on passing to the quotient gives an embedding of $\hat{A}/\sigma$ into $\mathbb{P}^5$; the image X is a complete intersection of 3 quadrics in $\mathbb{P}^5$. Let P denote the (2-dimensional) linear system of quadrics of $\mathbb{P}^5$ through X, and $\Delta \subset P$ the locus of singular quadrics; then show that Δ is the union of 6 lines of $P \cong \mathbb{P}^2$ tangent to a common conic, and that the double cover of the conic branched in the 6 points of tangency is isomorphic to C.

(Observe that $|4\Theta|$ contains certain divisors of the form $\sum_{i=1}^4 \Theta_{\alpha_i}$, where $\alpha_i \in JC_2$ satisfy $\sum_i \alpha_i = 0$.)

(10) Show that for any $g \geqslant 2$ there exists a $K3$ surface S and a curve $C \subset S$ such that the morphism defined by $|C|$ is a 2-to-1 cover of a rational surface in $\mathbb{P}^g$.
(For $g = 2k+1$, consider the double cover $\pi : S \to \mathbb{P}^1 \times \mathbb{P}^1$ branched along a smooth curve of bidegree $(4,4)$, and take $C \in |h_1 + kh_2|$ in the notation of I.9(b). For $g = 2k$, the argument is similar, considering a suitable double cover of $\mathbb{F}_1$, the blow-up of $\mathbb{P}^2$ in one point.)

(11) Let $S \subset \mathbb{P}^5$ be a $K3$ surface of degree 8 which is not a complete intersection of 3 quadrics. Show that there are 6 linear forms A_i, B_j on $\mathbb{P}^5$ ($1 \leqslant i, j \leqslant 3$) such that the linear system of quadrics containing S is given by the quadratic forms

$$\det \begin{pmatrix} A_1 & A_2 & A_3 \\ B_1 & B_2 & B_3 \\ \lambda & \mu & \nu \end{pmatrix}, \qquad (\lambda, \mu, \nu) \in \mathbb{P}^2.$$

Show that there exist 3 quadratic forms P, Q, R on $\mathbb{P}^5$ such that S has equations

$$A_iB_j - A_jB_i = 0 \qquad (\text{for } 0 \leqslant i < j \leqslant 3),$$
$$A_1P + A_2Q + A_3R = B_1P + B_2Q + B_3R = 0.$$

Deduce from this that S is a specialization of a complete intersection of 3 quadrics of $\mathbb{P}^5$. Let C be a smooth hyperplane section of S; show that C has a g_3^1, that is, an invertible sheaf L with $\deg L = 3$, $h^0(L) = 2$. Conversely, if a hyperplane section C of a $K3$ surface $S \subset \mathbb{P}^5$ of degree 8 has a g_3^1, then S is not a complete intersection of 3 quadrics.

(12) Let S be a $K3$ surface. Show that $H_1(S, \mathbb{Z}) = H^2(S, \mathbb{Z})_{\text{tors}} = 0$. Let X be an Enriques surface; show that $H^2(X, \mathbb{Z})_{\text{tors}}$ is the group of order 2 generated by $[K]$ (and therefore $H_1(X, \mathbb{Z}) \cong \mathbb{Z}/2$).

(13) Construct an irreducible variety T (resp. T') and a smooth morphism $X \to T$ (resp. $X' \to T'$) having fibres which are Enriques surfaces of the type constructed in Example VIII.18 (resp. VIII.19), and in such a way that every surface of the type considered is isomorphic to a finite (non-zero) number of fibres. Show that $\dim T = 10$, $\dim T' = 9$.

(14) In the notation of Example VIII.19, consider the varieties

$$\Delta = \{Q \in P \mid Q \text{ is singular }\} \subset P$$
$$W = \{x \in \mathbb{P}^3 \mid \exists Q \in P \text{ such that } x \in \operatorname{Sing} Q\} \subset \mathbb{P}^4.$$

Show that Δ is a quartic with 10 ordinary double points (*quartic symmetroid*), and that W is birationally equivalent to Δ; the projection of $X \subset \mathbb{P}^3 \times \mathbb{P}^3$ to one of its factors $\mathbb{P}^3$ defines a morphism $X \to W$ which is an isomorphism above a point $p \in W$ if and only if there is only one quadric of P singular at p. In particular, W is smooth if and only if Δ contains no line.

(15) Let S' be a surface of degree 6 having the 6 edges of a tetrahedron as double curves, and non-singular outside these edges; let $\pi : S \to S'$ be the normalization of S'. Show that S is an Enriques surface.

(Show that for $S' \subset \mathbb{P}^3$ a surface of degree d with 'ordinary singularities', that is a double curve C and triple points on C, the canonical divisors of the normalization are cut out by surfaces of degree $d-4$ passing through C. Deduce from this that $p_g(S) = 0$, $2K_S \equiv 0$; if ℓ_1 and ℓ_2 are two opposite edges of the tetrahedron, and $\tilde{\ell}_i = \pi^{-1}(\ell_i)$, show that the $\tilde{\ell}_i$ are elliptic curves and $K \equiv \tilde{\ell}_1 - \tilde{\ell}_2$. To prove that $q(S) = 0$, one can prove that $b_2 \geqslant 3$, and then use Noether's formula; for example, if ℓ, ℓ' and ℓ'' are three edges of a face of the tetrahedron, show that $\tilde{\ell}$, $\tilde{\ell}'$ and $\tilde{\ell}''$ are linearly independent in $NS(S)$.)

Show that the surface S depends on 10 moduli.

(16) Let C be a smooth sextic curve in $\mathbb{P}^2$, and S the double cover of $\mathbb{P}^2$ branched in C; let $H \subset S$ be the inverse image of a line of $\mathbb{P}^2$. Show that a curve of S not belonging to any linear system $|dH|$ is projected to a curve of $\mathbb{P}^2$ of degree n which touches C in $3n$ points. Deduce that for a sufficiently general choice of C, $\operatorname{Pic} S = \mathbb{Z}[H]$.

IX

SURFACES WITH $\kappa = 1$ AND ELLIPTIC SURFACES

Lemma IX.1 *Let S be a non-ruled minimal surface.*

(a) *If $K^2 > 0$, there exists an integer n_0 such that ϕ_{nK} maps S birationally onto its image for all $n \geqslant n_0$.*

(b) *If $K^2 = 0$ and $P_r \geqslant 2$, write $rK \equiv Z + M$, where Z is the fixed part of the system $|rK|$ and M is the mobile part. Then*

$$K.Z = K.M = Z^2 = Z.M = M^2 = 0 .$$

Proof

(a) Let H be a hyperplane section of S (for an arbitrary embedding). Since $K^2 > 0$, the Riemann–Roch theorem gives:

$$h^0(nK - H) + h^0(H + (1-n)K) \to \infty \quad \text{as} \quad n \to \infty.$$

We have $H.K > 0$, as S is non-ruled, hence $(H+(1-n)K).H < 0$ for n sufficiently large; this implies that $h^0(H+(1-n)K)$ is zero for large n. It follows that there exists n_0 such that $h^0(nK-H) \geqslant 1$ for $n \geqslant n_0$. Let $E \in |nK - H|$; it is clear that the system $|nK| = |H+E|$ separates points of $S-E$, and separates tangents to points of $S - E$. The restriction of ϕ_{nK} to $S - E$ is thus an embedding, which proves (a).

(b) Since $rK^2 = K.Z + K.M = 0$ and both $K.Z$ and $K.M$ are non-negative, we see that $K.Z = K.M = 0$. Since M is mobile, $Z.M$ and M^2 are also non-negative; from the equation $rK.M = Z.M + M^2 = 0$ we deduce that $Z.M = M^2 = 0$, and finally that $Z^2 = (rK - M)^2 = 0$.

Proposition IX.2 *Let S be a minimal surface with $\kappa = 1$.*

(a) *We have $K^2 = 0$.*

(b) *There is a smooth curve B and a surjective morphism $p : S \to B$ whose generic fibre is an elliptic curve.*

Proof By Lemma IX.1(a) we have $K^2 \leqslant 0$; so $K^2 = 0$ since otherwise S would be ruled (VI.2). Let r be an integer such that $P_r \geqslant 2$. Write Z for the fixed part of the system $|rK|$, M for the mobile part, so that $rK \equiv Z + M$. Lemma IX.1(b) gives $M^2 = K.M = 0$. It follows that $|M|$ defines a morphism f from S to $\mathbb{P}^N$ whose image is a curve C. Consider the Stein factorization (V.17) $f : S \xrightarrow{p} B \to C \subset \mathbb{P}^N$, where p has connected fibres. Let F be a fibre of p; since M is a sum of fibres of p and $K.M = 0$, we must have $K.F = 0$. It follows that $g(F) = 1$, so that the smooth fibres of p are elliptic curves.

A surface satisfying (b) in the proposition is called an elliptic surface; the proposition says that all surfaces with $\kappa = 1$ are elliptic. The converse is false, but we can say:

Proposition IX.3 *Let S be a minimal elliptic surface, $p : S \to B$ the elliptic fibration; for $b \in B$, put $F_b = p^*[b]$.*

(a) *We have $K_S^2 = 0$.*
(b) *S is either ruled over an elliptic curve, or a surface with $\kappa = 0$, or a surface with $\kappa = 1$.*
(c) *If $\kappa(S) = 1$, there exists an integer $d > 0$ such that*

$$dK \equiv \sum_i n_i F_{b_i}, \qquad n_i \in \mathbb{N},\ b_i \in B .$$

For r sufficiently large, the system $|rdK|$ is without base points and defines a morphism from S to $\mathbb{P}^N$ which factorizes as $f : S \xrightarrow{p} B \xrightarrow{j} \mathbb{P}^N$, where j is an embedding of B in $\mathbb{P}^N$.

Proof If S is ruled over a curve C, the elliptic curves F_b must be mapped surjectively onto C, which implies that C is either rational or elliptic; it follows that $K^2 \geqslant 0$ (III.21). Thus $K^2 \geqslant 0$ for all minimal elliptic surfaces (VI.2).

Suppose there exists $n \in \mathbb{Z}$ such that the system $|nK|$ is non-empty; let $D \in |nK|$. Since $D.F_b = 0$ for all b, the components of D are contained in the fibres of p; since $K^2 \geqslant 0$, Proposition VII.4 shows that:

$$D = \sum_i r_i F_{b_i} \qquad \text{for some } r_i \in \mathbb{Q},\ r_i \geqslant 0 .$$

It follows that $K^2 = D^2 = 0$ in this case.

Now let X be a minimal elliptic surface with $K_X^2 > 0$. The preceding argument shows that we must have $|nK_X| = \emptyset$ for all $n \in \mathbb{Z}$, $n \neq 0$. Then by Riemann–Roch $h^0(nK) + h^0((1-n)K_X) \to \infty$ as $n \to \infty$; this gives a contradiction, proving (a).

Let S be a minimal elliptic surface. Since $K.F_b = 0$, the maps ϕ_{nK} contract the fibres F_b; it follows that the image of ϕ_{nK} has dimension $\leqslant 1$, and hence that $\kappa = 0$, 1 or $-\infty$. This proves (b).

Finally if $\kappa = 1$, let us choose an integer n such that $P_n \geqslant 1$, and let $D \in |nK|$; then

$$D = \sum_i r_i F_{b_i} \qquad \text{for some } r_i \in \mathbb{Q},\ r_i \geqslant 0 .$$

Write $r_i = \dfrac{n_i}{m}$ (with $n_i, m \in \mathbb{N}$) and put $d = mr$; then

$$dK \equiv \sum n_i F_{b_i} = p^* A \qquad \text{where } A = \sum_i n_i [b_i] .$$

For r sufficiently large, the system $|rA|$ on B is without base points and defines an embedding j of B in $\mathbb{P}^N$; it follows that the system $|rdK| = p^*|rA|$ is without base points and defines the morphism $f = j \circ p$ of S to $\mathbb{P}^N$. This proves (c).

Examples IX.4 First we give some examples of elliptic surfaces with $\kappa \neq 1$.

(1) The ruled surface $E \times \mathbb{P}^1$ is obviously elliptic; more generally, if G is a group of translations of E acting on $\mathbb{P}^1$, the ruled surface $(E \times \mathbb{P}^1)/G$ is elliptic (cf. Exercise 3).

(2) The bielliptic surfaces are obviously elliptic: in fact they admit two distinct elliptic fibrations.

(3) An Abelian surface A is an elliptic surface if and only if there is an exact sequence $0 \to E \to A \to F \to 0$, where E and F are elliptic curves.

(4) In the course of proving VIII.15 we saw some elliptic $K3$ surfaces: quartics containing a line, complete intersections $(2,3)$ in $\mathbb{P}^4$ where the quadric is singular, etc. One can show that in the variety T_g parametrizing $K3$ surfaces of degree $(2g-2)$ in $\mathbb{P}^g$, the space of elliptic surfaces is a divisor (very reducible!).

(5) Enriques showed that every Enriques surface is elliptic (Exercise 7). For now we will simply check this for one example.

Consider the surface X/σ, where X is the $K3$ surface defined in $\mathbb{P}^5$ by the equations

$$P_i = Q_i(X_0, X_1, X_2) + Q'_i(X_3, X_4, X_5) = 0 \qquad (1 \leqslant i \leqslant 3),$$

and σ is the involution

$$(X_0, X_1, X_2, X_3, X_4, X_5) \mapsto (X_0, X_1, X_2, -X_3, -X_4, -X_5)$$

(see VIII.18). There are 9 points $(\lambda, \mu, \nu) \in \mathbb{P}^2$ for which the conics $\lambda Q_1 + \mu Q_2 + \nu Q_3$ and $\lambda Q'_1 + \mu Q'_2 + \nu Q'_3$ are both singular; the corresponding quadrics $\lambda P_1 + \mu P_2 + \nu P_3$ in $\mathbb{P}^5$ are therefore of rank 4, and so contain two pencils of 3-planes. Suppose for example that P_1 is one of the quadrics of rank 4; we choose a pencil $(L_t)_{t \in \mathbb{P}^1}$ of 3-planes contained in P_1. Since $L_t \cap X = L_t \cap P_2 \cap P_3$, the pencil (L_t) cuts a pencil on X of curves (C_t) which are complete intersections of two quadrics in $\mathbb{P}^3$; it follows that the generic curve C_t is elliptic. The involution σ preserves the pencils L_t and C_t (i.e. we have $\sigma(C_t) = C_{t'}$, with $t' \in \mathbb{P}^1$); the projection of the pencil (C_t) onto the Enriques surface X/σ defines a pencil of elliptic curves on this surface.

It is very easy to construct surfaces with $\kappa = 1$, and we end by giving an example:

Example IX.5 Let B be a smooth curve, p, q the two projections of $B \times \mathbb{P}^2$ onto B and $\mathbb{P}^2$, $|D|$ a base-point free linear system on B. A general divisor $S \in |p^*\mathcal{O}_B(D) \otimes q^*\mathcal{O}_{\mathbb{P}}(3)|$ is smooth; the restriction $p : S \to B$ is a fibration by plane cubics, hence elliptic curves. We have $K_S \equiv p^*(K_B + D)$, and when $\deg(D) > 2 - 2g(B)$ we have $\kappa(S) = 1$.

Historical Note IX.6

M. Noether noticed very early on that the canonical system of a surface can be without base points but composed of a pencil, which is then necessarily elliptic ([N1]). A detailed study of elliptic surfaces appears in [E7]; we have not dealt here with the principal results – the existence of the associated Jacobian fibration and the structure of the canonical divisor – because their rigorous justification is quite delicate. The theory was made watertight by Kodaira ([K]).

We remark that the study of elliptic surfaces can form the basis for the classification of surfaces: this is the point of view adopted by Bombieri and Mumford (in characteristic p, see [B-M]).

Exercises IX.7

(1) Let $p : S \to B$ be an elliptic fibration, $P : \mathrm{Alb}(S) \to JB$ the corresponding morphism. Show that P is surjective, and
either P is an isomorphism;
or its kernel is an elliptic curve E such that all smooth fibres of p are isogeneous to E.
In particular we have $q(S) = g(B)$ or $g(B) + 1$. (Observe that the image of F_b in $\mathrm{Alb}(S)$ is independent of b.)

(2) Let S be an elliptic surface, $p : S \to B$ the corresponding fibration, $F = \sum n_i C_i$ a reducible fibre of p. Let M' be the subgroup of $\mathrm{Pic}\, S$ generated by the C_i, so that $M = M'/\mathbb{Z}[F]$ carries a quadratic form induced by the intersection form. Show that the elements $r \in M$ with $r^2 = -2$ form a root system in M, of type A_n, D_n, or E_n, in which the C_i are the simple roots (cf. Bourbaki, *Groupes et Algèbres de Lie*, Chapter VI). Find the possible configurations for F.

(3) Let S be a ruled surface over an elliptic curve E; suppose that S is also an elliptic surface. Show that $S = (E \times \mathbb{P}^1)/G$, where G is a group of translations of E acting on $\mathbb{P}^1$.

(4) Let S be an Enriques surface, E an elliptic curve on S. Show that
either $h^0(E) = 1$; and then $|2E|$ is a pencil of elliptic curves without base points;
or $|E|$ is a base-point free pencil of elliptic curves; the corresponding fibration has exactly two multiple fibres E_1 and E_2, with $E \equiv 2E_1 \equiv 2E_2$, and $E_1 - E_2 \equiv K$.
(Notice that $|K + E| \neq \emptyset$ by Riemann–Roch; if $h^0(E) = 1$ then for $E' \in |K + E|$ we have $E \cap E' = \emptyset$, so that Riemann–Roch gives $h^1(-E - E') \geqslant 1$ and $h^0(2E) \geqslant 2$; use Exercise 12 from Chapter VIII.)

(5) Let S be the Enriques surface associated to the Reye congruence (Example VIII.19). Show that S is an elliptic surface.
(Show that the system of quadrics P contains 10 quadrics of rank 2. Such a quadric is the union of two planes L_1, L_2; the system P induces a net of conics on L_1. Show that the variety of lines contained in a conic of this net is an elliptic curve, which is naturally embedded in S; use Exercise 4.)

(6) Let S be a $K3$ surface. Show that if a divisor D on S satisfies $D^2 = 0$ and $D.C \geqslant 0$ for all smooth rational curves C, then

$D \equiv kE$ where E is an elliptic curve (and $k \geqslant 1$). Deduce that a $K3$ surface is elliptic if and only if its Picard group contains a non-zero element whose square is zero.
(If $D \equiv Z + M$, where Z is the fixed part of $|D|$, show that $D.Z = Z.M = M^2 = 0$, whence $Z^2 = 0$ which implies $Z = 0$. If C is a smooth rational curve, write s_C for the symmetry of Pic S defined by $s_C(D) = D + (C.D)C$; let W be the group generated by all the s_C. Show that for every divisor D with $D^2 = 0$, there exists $w \in W$ such that $C.w(D) \geqslant 0$ for all smooth rational curves C.)

(7) Show that every Enriques surface S is elliptic.
(Using the properties of quadratic forms, show that there exists a divisor D on S such that $D^2 = 0$. Deduce from Exercise 6 that the canonical double cover of S is an elliptic $K3$ surface; conclude that S is elliptic.)

X
SURFACES OF GENERAL TYPE

Proposition X.1 *Let S be a minimal surface; then the following 3 conditions are equivalent:*

(i) $\kappa(S) = 2$;
(ii) $K_S^2 > 0$ *and S is irrational;*
(iii) *there exists an integer n_0 such that ϕ_{nK} is a birational map of S to its image for $n \geqslant n_0$.*

If these conditions hold, S is called a surface of general type.

Proof Obviously, (iii) $\Rightarrow$ (i). Let S be a surface with $K^2 = 0$; then for any $n \geqslant 1$ the mobile part M of $|nK|$ satisfies $M^2 = 0$ (by Lemma IX.1(b)), which implies that $\dim \phi_{nK}(S) \leqslant 1$; hence (i) $\Rightarrow$ (ii). Finally, an irrational surface with $K^2 > 0$ is non-ruled (III.21); Lemma IX.1(a) now shows that (ii) $\Rightarrow$ (iii).

Remark X.2 The statement in (iii) can be substantially improved: in fact, $|nK|$ is base-point free for $n \geqslant 4$; as soon as $n \geqslant 5$, the morphism ϕ_{nK} is an isomorphism outside certain rational curves, contracted onto a very simple type of singularity (rational double points). We refer to [Bo] for a detailed analysis.

Examples X.3 The only difficulty is choosing from the wealth of examples.

(1) All surface complete intersections, with the exceptions S_2, S_3,S_4, $S_{2,2}$, $S_{2,3}$, $S_{2,2,2}$, are of general type (VII.5).
(2) Any product of curves of genus $\geqslant 2$ (or more generally, any surface fibred over a curve of genus $\geqslant 2$ with generic fibre of genus $\geqslant 2$) is of general type (VII.4).

(3) If $f : S' \to S$ is a surjective morphism, and if S is a surface of general type, then so is S'; this follows at once from the fact that $K_{S'} \equiv \Delta + f^*K_S$, with $\Delta \geqslant 0$.

(4) It is interesting to note that there exist surfaces of general type with $p_g = q = 0$; here is an example, the Godeaux surface. Let $S' \subset \mathbb{P}^3$ be the quintic given by $X^5 + Y^5 + Z^5 + T^5 = 0$, and let σ be the automorphism of order 5 of S' defined by $\sigma(X, Y, Z, T) = (X, \zeta Y, \zeta^2 Z, \zeta^3 T)$, where ζ is a primitive 5th root of unity. The group $G \cong \mathbb{Z}/5$ generated by σ acts on S' without fixed points, and the quotient $S = S'/G$ is thus smooth. Since $q(S') = 0$ and $K_{S'} \equiv H$ (by VIII.9 and IV.11), we have $p_g(S') = 4$ and $K_{S'}^2 = 5$. It follows that $q(S) = 0$ and $\chi(\mathcal{O}_{S_1}) = \frac{1}{5}\chi(\mathcal{O}_{S'}) = 1$ (using VI.3), hence $p_g = 0$; furthermore, $K_S^2 = \frac{1}{5}K_{S'}^2 = 1$, so that S is a surface of general type with $p_g = q = 0$ and $K^2 = 1$.

For more examples of this kind, see Exercises 3 and 4.

In view of the diversity of these examples, one does not hope to describe completely all surfaces of general type. The natural questions turning up are of a more general nature: for example, finding the possible numerical invariants for such a surface. This problem is still some way from a complete solution. We finish this course with one important inequality, due to Castelnuovo.

Theorem X.4 *Let S be a non-ruled surface; then $\chi_{\text{top}}(S) \geqslant 0$ and $\chi(\mathcal{O}_S) \geqslant 0$. Moreover, if S is of general type, then $\chi(\mathcal{O}_S) > 0$.*

The inequalities for $\chi(\mathcal{O}_S)$ are of course a consequence of that for χ_{top}, together with Noether's formula. Although we do not require this, it can be noted that the result follows from the classification for surfaces with $\kappa = 0$, and from Proposition X.10 below for elliptic surfaces, so that X.4 is actually a theorem on surfaces of general type.

We will require the following two propositions:

Fact X.5 *Let S be a surface; then*

(i) $b_2 \geqslant 2p_g$;

(ii) *if ω is a holomorphic* 1*-form on S, $d\omega = 0$.*

The first statement comes from Hodge theory, which gives the more precise formula $b_2 = 2p_g + h^1(S, \Omega_S^1)$. The second is much more elementary: Stokes' formula shows that

$$\int_S d\omega \wedge (\overline{d\omega}) = \int_S d(\omega \wedge \overline{d\omega}) = 0.$$

As $d\omega$ is a $(2,0)$-form, this implies $d\omega = 0$.

Lemma X.6 *Let S be a surface with $\chi_{\text{top}}(S) < 0$; then S has an étale cover $S' \to S$ such that $p_g(S') \leqslant 2q(S') - 4$.*

Proof The inequality $\chi_{\text{top}}(S) = 2 - 4q + b_2 < 0$ implies $q(S) \geqslant 1$; hence there are connected étale covers $S' \to S$ of any degree m (compare V.14.(5)). Take $m \geqslant 6$; then $\chi_{\text{top}}(S') \leqslant -6$, so that in view of X.5(i) we have

$$2 - 4q + 2p_g \leqslant -6\ , \qquad \text{that is} \quad p_g \leqslant 2q - 4\ .$$

Lemma X.7 *Let V, W be two finite dimensional complex vector spaces, and $\phi : \wedge^2 V \to W$ a homomorphism. Then if $\dim W \leqslant 2\dim V - 4$, there exist two linearly independent vectors v, $v' \in V$ such that $\phi(v \wedge v') = 0$.*

Proof Let C denote the cone of $\wedge^2 V$ made up of vectors of the form $v \wedge v'$, and let $\overline{C}$ be its image in the projective space $\mathbb{P}(\wedge^2 V)$. Consider also the Grassmannian $G_2(V)$ of 2-planes of V; sending a 2-plane $P \subset V$ to the line $\wedge^2 P \subset \wedge^2 V$ gives a morphism f of $G_2(V)$ into $\mathbb{P}(\wedge^2 V)$ whose image is just $\overline{C}$. It is easy to check that f is an embedding (the 'Plücker embedding'), and $\dim G_2(V) = 2(\dim V - 2)$. Hence $\dim C = 2\dim V - 3$. The formula for the dimension of intersections gives

$$\dim(C \cap \operatorname{Ker} \phi) \geqslant \dim C - \operatorname{codim}\ \operatorname{Ker} \phi \geqslant 2\dim V - 3 - \dim W\ ;$$

using the hypothesis shows that this is strictly positive, so that $C \cap \operatorname{Ker} \phi \neq \{0\}$. This is the assertion of the lemma.

The lemma applies to a surface S by setting $V = H^0(S, \Omega_S^1)$ and $W = H^0(S, \Omega_S^2)$, and taking ϕ to be the cup product. We thus get:

Corollary X.8 *Let S be a surface for which $p_g \leqslant 2q - 4$; then there exist two 1-forms ω_1 and ω_2 which are linearly independent in $H^0(S, \Omega_S^1)$, and such that the 2-form $\omega_1 \wedge \omega_2$ is zero.*

The following result, due to Castelnuovo and de Franchis, is the main point of the proof of X.4.

Proposition X.9 *Let S be a surface having two linearly independent 1-forms ω_1, $\omega_2 \in H^0(S, \Omega^1_S)$ such that $\omega_1 \wedge \omega_2 = 0$. Then there is a smooth curve B of genus $g \geqslant 2$, a surjective morphism $p : S \to B$ with connected fibres, and two 1-forms α_1, $\alpha_2 \in H^0(B, \Omega^1_B)$ such that $\omega_1 = p^*\alpha_1$, $\omega_2 = p^*\alpha_2$.*

Proof Let $K(S)$ denote the rational function field of S. Since $\omega_1 \wedge \omega_2 = 0$, we can write $\omega_2 = g\omega_1$ for some $g \in K(S)$. Using X.5(ii) we have $d\omega_1 = d\omega_2 = 0$, giving $\omega_1 \wedge dg = 0$, whence $\omega_1 = f\,dg$, with $f \in K(S)$. Now $d\omega_1 = 0$ gives $df \wedge dg = 0$, which is to say that f and g are algebraically dependent: there is a polynomial P in 2 variables such that $P(f, g) = 0$. Consider the projective curve $C \subset \mathbb{P}^2$ given by the affine equation $P(x, y) = 0$; the functions f, g define a rational map $\phi : S \dashrightarrow C$. Using elimination of indeterminacy (II.7), there exists a composite of blow-ups $\epsilon : \hat{S} \to S$ such that $h = \phi \circ \epsilon$ is a morphism. Now using Stein factorization (V.17), we get

$$h : \hat{S} \xrightarrow{q} B \xrightarrow{u} C ,$$

where B is a smooth curve, and q has connected fibres. Set $\alpha_1 = u^*(x\,dy)$ and $\alpha_2 = u^*(xy\,dy)$. These are meromorphic forms on B which by construction satisfy $q^*\alpha_i = \epsilon^*\omega_i$, for $i = 1, 2$. An easy local calculation shows that $\mathrm{div}(\epsilon^*\omega_i) = q^*\mathrm{div}\alpha_i + \sum_{p \in B} (n_i - 1)C_i$, where the sum takes place over the non-reduced fibres $q^*p = \sum n_i C_i$; in particular, the α_i are in fact holomorphic on B. They are linearly independent over $\mathbb{C}$, so that we deduce $g(B) \geqslant 2$. But then the exceptional curves of ϵ in $\hat{S}$ are also contracted by q; this implies that $q \circ \epsilon^{-1} = p : S \to B$ is a morphism having all the properties claimed.

Proposition X.10 *Let S be a surface, and $p : S \to B$ a surjective morphism of S to a smooth curve; suppose that p has connected fibres, and F_η is the generic fibre. Then $\chi_{\mathrm{top}}(S) \geqslant \chi_{\mathrm{top}}(B)\,\chi_{\mathrm{top}}(F_\eta)$.*

Proof Recall that by Lemma VI.4 we have the formula

$$\chi_{\mathrm{top}}(S) = \chi_{\mathrm{top}}(B)\,\chi_{\mathrm{top}}(F_\eta) + \sum_{s \in \Sigma} (\chi_{\mathrm{top}}(F_s) - \chi_{\mathrm{top}}(F_\eta)) ,$$

where Σ is the set of points of B for which the fibre F_s is singular. It is thus enough to show that $\chi_{\mathrm{top}}(F_s) \geqslant \chi_{\mathrm{top}}(F_\eta)$ for $s \in \Sigma$. Note that since χ_{top} increases on making a blow-up, we can suppose that the fibres of p do not contain any exceptional curves.

Suppose first that $F_s = nC$, with $n \geqslant 1$ and C irreducible. Using the inequality $\chi_{\text{top}}(C) \geqslant 2\chi(\mathcal{O}_C)$ of Lemma VI.5, we get

$$\chi_{\text{top}}(F_s) = \chi_{\text{top}}(C) \geqslant -C.K = -\frac{1}{n}F.K = \frac{1}{n}\chi_{\text{top}}(F_\eta) \, .$$

If $g(F_\eta) \geqslant 1$, this gives the required inequality $\chi_{\text{top}}(F_s) \geqslant \chi_{\text{top}}(F_\eta)$. The only other case is when $F_\eta \cong \mathbb{P}^1$; then $C.K = -\frac{2}{n}$ and $C^2 = 0$ imply $n = 1$ and $C \cong \mathbb{P}^1$, giving $\chi_{\text{top}}(F_s) = \chi_{\text{top}}(F_\eta)$.

Now suppose that F_s is reducible, $F_s = \sum n_i C_i$: then

$$\chi_{\text{top}}(F_s) = \chi_{\text{top}}\left(\sum C_i\right) \geqslant -\left(\sum C_i\right)^2 - \left(\sum C_i\right).K \, .$$

Corollary VIII.4 shows that $(\sum C_i)^2 \leqslant 0$ and $C_i^2 < 0$ for each i; since C_i is not exceptional, $C_i.K \geqslant 0$, and hence

$$\chi_{\text{top}}(F_s) \geqslant -\sum n_i(C_i.K) = -F_s.K = -F_\eta.K = \chi_{\text{top}}(F_\eta) \, ,$$

giving the result.

Proof of Theorem X.4 Let S be a surface with $\chi_{\text{top}}(S) < 0$; Lemma X.6 shows that there is an étale cover $S' \to S$ having $\chi_{\text{top}}(S') < 0$, $p_g(S') \leqslant 2q(S') - 4$. Corollary X.8 and Proposition X.9 show that S' has a surjective morphism with connected fibres $p : S' \to B$ with $g(B) \geqslant 2$. If S' is ruled, then so is S, which is excluded by hypothesis; hence $g(F_\eta) \geqslant 1$. Proposition X.10 then gives $\chi_{\text{top}}(S') \geqslant 0$, which is a contradiction.

Remark X.11 In view of Noether's formula, $\chi_{\text{top}}(S) \geqslant 0$ is equivalent to the inequality $K_S^2 \leqslant 12\chi(\mathcal{O}_S)$. For a non-ruled surface, the very much deeper inequality $K^2 \leqslant 9\chi(\mathcal{O}_S)$ was proved by Bogomolov and Miyaoka, and independently by S.T. Yau.

Historical Note X.12

As we have noted, both the classical and the modern results on surfaces of general type are substantially less complete than those for surfaces with $\kappa \leqslant 1$. Among the classical results are the inequalities of Noether ([N1]), see also Exercise 1) and of Castelnuovo ([C3]); the structure of the maps ϕ_{nK} was tackled by Enriques in [E8]. He proved notably (restricting himself to surfaces with $q = 0$) that if $p_g \geqslant 1$, ϕ_{2K} is generically finite to a surface (see Exercise 6), and that ϕ_{3K} is a birational embedding for all but a small number of exceptional surfaces – although his proof of this last point is incomplete.

Questions on the pluricanonical maps ϕ_{nK} for $n \geqslant 2$ were definitely settled by Kodaira and Bombieri (see [Bo]); the structure of ϕ_K is studied in [B1]. Among more recent results one should also mention Bogomolov's work on the cotangent bundle Ω^1, and the related inequality $K^2 \leqslant 9\chi(\mathcal{O})$ mentioned in X.11.

Exercises X.13

(1) Let S be a minimal surface of general type. Prove that $K^2 \geqslant 2p_g - 4$.
(Split into cases according as to whether the image of ϕ_K is a curve or a surface; in the second case, apply Clifford's theorem on canonical curves – compare Chapter VI, Exercise 2.)

(2) Let S be a minimal surface of general type with $p_g = q = 0$ and $K^2 = 1$. Show that $H_1(S, \mathbb{Z})$ has order $\leqslant 5$; if the order is 5, S is a Godeaux surface.
(If S^{ab} is the Abelian universal cover of S, which is Galois with group $H_1(S, \mathbb{Z})$, show by using Exercise 1 that the order of the cover $S^{ab} \to S$ is $\leqslant 5$. If equality holds, show that ϕ_K is a birational morphism from S^{ab} to a quintic of $\mathbb{P}^3$.)

(3) Let $G = (\mathbb{Z}/2)^3$. Find an action of G on $\mathbb{P}^6$ and a surface $S' \subset \mathbb{P}^6$, the intersection of 4 quadrics invariant under G, in such a way that G acts freely on S'. Show that the quotient surface $S = S'/G$ satisfies $p_g = q = 0$, $K^2 = 2$.

(4) Let $C \subset \mathbb{P}^2$ be the plane quintic $X^5 + Y^5 + Z^5 = 0$; the group $G = (\mathbb{Z}/5)^2$ acts on C by $(a, b) \cdot (X, Y, Z) = (\zeta^a X, \zeta^b Y, Z)$. Show that for a suitable automorphism ϕ of G, the action of G on $C \times C$ given by $g.(x, y) = (gx, \phi(g)y)$ is fixed-point free. The surface $S = (C \times C)/G$ then satisfies $p_g = q = 0$, $K^2 = 8$. Find more such examples.

(5) Let S be minimal surface of general type with $K^2 = 1$. Show that $q = 0$, $p_g \leqslant 2$; find examples with $p_g = 0, 1, 2$. (Use Exercise 1.)

(6) Let S be a surface of general type with $q = 0$, $p_g \geqslant 1$; show that the image of ϕ_{2K} is a surface.
(If $2K \equiv Z + aF$, where Z is the fixed part of $|2K|$, show that $|K + F|$ induces the complete canonical system on F.)

APPENDIX A: CHARACTERISTIC p

The classification of surfaces was extended to the case of an arbitrary algebraically closed base field by Bombieri and Mumford (some results having been obtained previously by Zariski). It follows from their work that the classification of surfaces in characteristics $\neq 2, 3$ is identical to that over $\mathbb{C}$; in characteristics 2 and 3 certain 'non-classical' surfaces appear. We restrict ourselves to stating the results, referring to [B-M] for the proofs.

(1) All the theorems on ruled and rational surfaces are true in all characteristics (Noether–Enriques, Castelnuovo, structure of minimal models, ...). In particular the ruled surfaces are characterized by the condition $P_{12} = 0$, or by the termination of adjunction, or by the inequality $C.K < 0$ for some non-exceptional curve C.

(2) The list of surfaces with $\kappa = 0$ comprises:

(a) the surfaces with $K \equiv 0$, $q = 0$ ($q = h^1(\mathcal{O}_S)$), called $K3$ surfaces; these surfaces have the same properties as in characteristic 0 (VIII.8 to VIII.16). Furthermore, Deligne has shown that these surfaces are obtained from $K3$ surfaces in characteristic 0 by reduction mod p;

(b) Abelian surfaces;

(c) surfaces with $2K \equiv 0$, $p_g = q = 0$, called (classical) Enriques surfaces. In characteristic $\neq 2$, these are still quotients of $K3$ surfaces by fixed-point free involutions, and have all the properties of Enriques surfaces over $\mathbb{C}$; in particular they are elliptic. In characteristic 2 the canonical double cover is inseparable, with group μ_2 ([M3]); the surfaces are elliptic or quasi-elliptic (a surface S is quasi-elliptic if it has a morphism to a smooth curve whose generic fibre is a cuspidal cubic);

(d) bielliptic surfaces, of the form $(E \times F)/G$. In characteristic $\neq 2, 3$, we obtain all of the types 1 to 7 listed in VI.20; in characteristic 3, type 6 ($G = (\mathbb{Z}/3)^3$) does not exist; in characteristic 2, type 4 ($G = \mathbb{Z}/4 \times \mathbb{Z}/2$) does not exist, and for type 2 we must write $G = \mathbb{Z}/2 \times \mu_2$;

(e) in characteristic 2, surfaces with $K \equiv 0$, $q = 1$, which are called non-classical Enriques surfaces. They fall into two classes, the 'singular' (where the Frobenius automorphism of $H^1(\mathcal{O}_S)$ is bijective) and the 'supersingular' (where it is zero). The first are quotients of $K3$ surfaces by a fixed-point free involution, the others have a canonical inseparable cover with group α_2. All these surfaces are elliptic or quasi-elliptic;

(f) in characteristic 2 or 3, 'quasi-bielliptic' surfaces of the form $(E \times C)/G$ where E is an elliptic curve, C a cuspidal cubic, G a group of translations of E operating on C. One can set up a list, analogous to VI.20, of all the possible cases. These surfaces satisfy $4K \equiv 0$ or $6K \equiv 0$.

(3) The surfaces with $\kappa = 1$ are elliptic or possibly (in characteristic 2 or 3) quasi-elliptic.

(4) Among the irrational surfaces, the surfaces with $\kappa = 2$ are characterized by $K^2 > 0$. The structure of the pluricanonical maps ϕ_{nK} ($n \geqslant 2$) has been studied in [Ek]; most of the results of Bombieri and Kodaira still hold. On the other hand, Castelnuovo's inequality does not always hold.

APPENDIX B: COMPLEX SURFACES

Compact complex surfaces can be classified according to their Kodaira dimension, more or less along the lines we have followed in this book. I refer to the original papers of Kodaira or to the book [B-P-V] for a thorough treatment. Here I would like to give a general idea and point out a few new phenomena which occur. It is actually a good exercise for the reader to go back through this book and find out for which arguments the algebraicity is really needed.

One of the first problems we meet on our way is about numerical invariants: Hodge theory does not hold in general, so we cannot expect the equalities

$$q(X) = \frac{1}{2}b_1(X) = \dim H^0(X, \Omega^1_X) = \dim H^1(X, \mathcal{O}_X)$$

to hold for any compact complex surface X. Recall that we defined $q(X)$ as $\dim H^1(X, \mathcal{O}_X)$; let us put $h := \dim H^0(X, \Omega^1_X)$. An important role will also be played by the number b^+ (resp. b^-) of positive (resp. negative) eigenvalues of the intersection form on $H^2(X, \mathbb{R})$.

Proposition B.1 *Let X be a compact complex surface. Then either $b_1(X)$ is even, $b^+ = 2p_g + 1$, and $q(X) = h(X) = \frac{1}{2}b_1(X)$; or $b_1(X)$ is odd, $b^+ = 2p_g$, $h(X) = q(X) - 1$ and $b_1(X) = 2q(X) - 1$.*

Sketch of proof The key point is the signature formula (see the Historical Note of Chapter I):

$$b^+ - b^- = \frac{1}{3}(K^2_X - 2\chi_{\text{top}}(X)) .$$

Comparing with Noether's formula I.14 gives

$$(4 - 4q + 4p_g) - (b^+ - b^-) = \chi_{\text{top}}(X) = 2 - 2b_1 + b_2 ,$$

hence, since $b_2 = b^+ + b^-$,

$$(b^+ - 2p_g) + (2q - b_1) = 1 \ . \tag{1}$$

We claim that both terms appearing in the left hand side are non-negative. For the first one, this is because $H^0(X, K_X) \oplus \overline{H^0(X, K_X)}$ embeds into $H^2(X, \mathbb{C})$ (viewed as de Rham cohomology), and the restriction of the intersection form to the real part of the image is easily seen to be positive. For the second, one uses the exact sequence

$$0 \to \mathbb{C} \to \mathcal{O}_X \xrightarrow{d} \mathcal{Z}\Omega^1_X \to 0 \ ,$$

where $\mathcal{Z}\Omega^1_X$ is the sheaf of closed holomorphic 1-forms. By X.5(ii), every global holomorphic 1-form is closed, so we have an exact sequence

$$0 \to H^0(X, \Omega^1_X) \to H^1(X, \mathbb{C}) \to H^1(X, \mathcal{O}_X) \ ,$$

hence $b_1 \leqslant h + q$. On the other hand one can see as for 2-forms that $H^0(X, \Omega^1_X) \oplus \overline{H^0(X, \Omega^1_X)}$ embeds into $H^1(X, \mathbb{C})$; this gives $2h \leqslant b_1$, thus $h \leqslant q$, and finally

$$2h \leqslant b_1 \leqslant h + q \leqslant 2q \ , \tag{2}$$

which implies in particular $2q - b_1 \geqslant 0$.

Thus equation (1) leaves two possibilities:
either $b^+ = 2p_g + 1$ and $b_1 = 2q$, which in view of (2) implies $h = q$;
or $b^+ = 2p_g$ and $b_1 = 2q - 1$, which gives $h = q - 1$.

So we see that compact complex surfaces have a rather different behaviour according to the parity of b_1.

Surfaces with b_1 even

Numerically, these surfaces behave like algebraic surfaces; in fact it is now known that they are *Kähler* surfaces, so that Hodge theory can be applied. We find the same classification as in the algebraic case, essentially with the same arguments. Note that some of these surfaces are automatically algebraic because of the following result, due to Kodaira:

Proposition B.2 *A compact complex surface which admits a line bundle L with $c_1(L)^2 > 0$ is projective.*

As a consequence, every surface with $p_g = 0$ and b_1 even is algebraic: Proposition 1 provides a cohomology class x with $x^2 > 0$, and the exponential exact sequence tells us that x is the first Chern class of some line bundle. Therefore complex minimal surfaces with b_1 even fall into the following classes:

$\kappa = -\infty$: rational and ruled surfaces (algebraic).

$\kappa = 0$: complex tori and $K3$ surfaces (not necessarily algebraic), Enriques and bielliptic (algebraic).

$\kappa = 1$: elliptic surfaces.

$\kappa = 2$: surfaces of general type (algebraic).

Surfaces with b_1 odd

This is where we get really new surfaces. The case $\kappa = 2$ cannot occur, by Propositions B.2 and X.1. The case $\kappa = 1$ gives non-algebraic elliptic surfaces. So the two interesting cases are $\kappa = -\infty$ and $\kappa = 0$.

$\kappa = -\infty$: this is the famous class VII_0 of Kodaira. It is not difficult to show that surfaces in this class have $b_1 = q = 1$. Surfaces of this type with the extra condition $b_2 = 0$ are now completely classified ('Bogomolov's theorem'): they are either *Hopf surfaces* (quotients of $\mathbb{C}^2 - \{0\}$ by a discrete group acting freely) or the so-called *Inoue surfaces*. Though various examples are known, the case $b_2 > 0$ is far less understood.

$\kappa = 0$: two (related) new types appear. The *primary Kodaira surfaces* can be obtained as follows. Let B be an elliptic curve, L a line bundle of degree $\neq 0$ on B, L^* the complement of the zero section in L. The group $\mathbb{C}^*$ acts on L^*; let $X = L^*/q^{\mathbb{Z}}$, where $q^{\mathbb{Z}}$ is an infinite cyclic subgroup of $\mathbb{C}^*$. The (compact) surface X is a principal bundle over B with structure group the elliptic curve $\mathbb{C}^*/q^{\mathbb{Z}}$. It follows easily that $K_X \equiv 0$, $q(X) = 2$, $b_1(X) = 3$.

Some primary Kodaira surfaces may admit a free action of a finite group; the quotient is called a *secondary Kodaira surface*. These surfaces have $q = b_1 = 1$, $K \not\equiv 0$ but $nK \equiv 0$ with $n = 2, 3, 4$ or 6.

APPENDIX C: FURTHER READING

With the exception of Appendix B, this book was written in 1978. Since then new results have been obtained, and other books have appeared, using somewhat different approaches to the classification. To describe these, let me first introduce a piece of terminology, due to M. Reid, which is now universally adopted: a divisor D on a variety X is said to be *nef* if $D.C \geqslant 0$ for all curves C in X. Now the crucial point of the classification lies in the following implications, for a minimal surface X:

$$\kappa(X) = -\infty \;\Rightarrow\; K_X \text{ not nef } \;\Rightarrow\; X \text{ is ruled} .$$

This is our Corollary VI.18 (the other implications in the Corollary are not difficult); we actually proved directly that a surface X with $\kappa = -\infty$ is ruled. The proof is easy for $q \geqslant 2$; we used Castelnuovo's theorem for $q = 0$, and a rather lengthy analysis in the case $q = 1$, proving first that the surface is the quotient of a product of two curves by a finite group, and then, by computing the numerical invariants, that it is ruled.

One can slightly shorten this approach by proving directly that the surface is elliptic, with rational base; then a computation of the canonical bundle does the job. This is the point of view taken in [G-H], and also (somewhat indirectly) in [B-M].

In [B-P-V], the authors use instead (a particular case of) *Iitaka's conjecture*: if X is a compact complex surface, $p : X \to B$ a fibration and F a smooth fibre of p, one has

$$\kappa(X) \geqslant \kappa(B) + \kappa(F) .$$

Proving this statement requires some hard work; the reward is an immediate proof of the fact that an algebraic surface with $\kappa = -\infty$ and $q \geqslant 1$ is ruled (apply the above inequality to the Albanese fibration).

I would also like to mention an elegant proof of the implication

$\{K_X \text{ not nef}\} \Rightarrow \{X \text{ ruled}\}$ for a minimal surface X, inspired by Mori theory. Here the key point is the ***rationality theorem***: if X is an algebraic surface such that K_X is not nef and H is an ample divisor on X, the number

$$b = \sup\{t \in \mathbb{R} \mid H + tK \text{ is nef}\}$$

is ***rational.*** It is not difficult to deduce from this statement (which can be proved by elementary means) that a minimal surface with K not nef is either $\mathbb{P}^2$ or geometrically ruled (see e.g. [P]). Unfortunately this nice method does not seem to give the first of the two implications we discussed above.

The theory of the period map for $K3$ surfaces (alluded to in the Historical Note of Chapter VIII) is now fully understood, thanks to the work of Looijenga, Todorov, Siu, ... It gives a nice description of the various moduli spaces of $K3$ surfaces as quotients of bounded domains by discrete groups. A short account can be found in [B2], and a detailed treatment in [X2] or [B-P-V]; this last reference includes applications to Enriques surfaces. An extensive treatment of Enriques surfaces can be found in the monograph [C-D].

The work on surfaces of general type has been oriented in two main directions: structure of the canonical ring and 'geography' (this means studying the possible values of the numerical invariants, particularly $\chi(\mathcal{O})$ and K^2, for a surface of general type, and the influence of these values on the geometry of the surface). Perhaps the most spectacular progress is the elegant ***Reider's method*** which simplifies and improves the results of Bombieri on pluricanonical maps. For an excellent overview (though not completely up-to-date by now) I recommend the lectures by F. Catanese and U. Persson at the Bowdoin AMS Summer Institute in 1985 (*Proc. of Symposia in Pure Math.* **46**, (1987)), as well as the brief survey [P].

Though it is beyond the scope of this book, I cannot resist mentioning the exciting events of these last 10 years about the topology of algebraic surfaces. The story started around 1985, when S. Donaldson proved that two algebraic surfaces (one elliptic, the other one rational), known to be homeomorphic by general principles, were not diffeomorphic. More generally, he produced, using the moduli space of rank 2 vector bundles on a surface (together with some hard analysis), a set of invariants of the C^∞-structure, the ***Donaldson polynomials.*** The computation of these invariants became a whole industry, culminating in 1994 with the proof by Friedman and Qin of the 'Van de Ven conjecture': ***the Kodaira dimension is a differentiable invariant*** [F-Q].

Shortly afterwards, E. Witten announced that a new set of invariants he had just constructed with N. Seiberg could be used instead of the Donaldson invariants [Wi]. No hard analysis is needed, and (unfortunately for us!) very little algebraic geometry. So one gets a relatively easy proof of the Van de Ven conjecture, and even of a stronger statement: for each $n \geqslant 1$, ***the plurigenus*** P_n ***is a differentiable invariant*** (see e.g. [F-M]).

REFERENCES

[A] A. ANDREOTTI *On the complex structures of a class of simply-connected manifolds.* In Algebraic Geometry and Topology (a symposium in honour of S. Lefschetz), Princeton University Press (1957).

[B1] A. BEAUVILLE *L'application canonique pour les surfaces de type général.* Invent. Math. **55** (1979).

[B2] A. BEAUVILLE *Surfaces K3.* Sém. Bourbaki, Exp. 609, Astérisque **105-106** (1983).

[B-DF] G. BAGNERA and M. DE FRANCHIS *Sopra le superficie algebriche che hanno le coordinate del punto generico esprimibili con funzioni meromorfe 4^{ente} periodiche di 2 parametri.* Rend. Acc. Lincei **16** (1907).

[B-M] E. BOMBIERI and D. MUMFORD *Enriques' classification of surfaces in char. p.* Part I in *Global Analysis*, Princeton University Press (1969); Part II in *Complex Analysis and Algebraic Geometry*, Cambridge University Press (1977); Part III in *Invent. Math.* **35** (1976).

[Bo] E. BOMBIERI *Canonical models of surfaces of general type.* Publ. Math. IHES **42** (1973).

[B-P-V] W. BARTH, C. PETERS and A. VAN DE VEN *Compact Complex Surfaces.* Springer-Verlag (1984).

[C1] G. CASTELNUOVO *Sulla razionalità delle involuzioni piane.* Math. Ann. **44** (1894).

[C2] G. CASTELNUOVO *Sulle superficie di genere zero.* Memorie della Società dei XL, s. III, **10** (1894–96).

[C3] G. CASTELNUOVO *Sulle superficie aventi il genere aritmetico negativo.* Rend. di Palermo, **20** (1905).

[C-D] F. COSSEC and I. DOLGACHEV *Enriques surfaces I.* Birkhäuser (1989).

[C-E] G. CASTELNUOVO and F. ENRIQUES *Sopra alcune questioni fondamentali nella teoria delle superficie algebriche.* Ann. di Mat. pura ed app., s. 3^a, **6** (1901).

[C-G] C.H. CLEMENS and P. GRIFFITHS *The intermediate Jacobian of the cubic threefold.* Ann. of Math. **95** (1972).

[DP] P. DEL PEZZO *Sulle superficie dell' n^{mo} ordine immerse nello spazio a n dimensioni.* Rend. di Palermo, **1** (1887).

[E1] F. ENRIQUES *Introduzione alla geometria sopra le superficie algebriche.* Memorie della Società dei XL, s. 3^a, **10** (1896).

[E2] F. ENRIQUES *Sopra le superficie che posseggono un fascio di curve razionali.* Rend. Acc. Lincei, s. 5^a, **7** (1898).

[E3] F. ENRIQUES *Sulle superficie algebriche di genere geometrico zero.* Rend. di Palermo, **20** (1905).

[E4] F. ENRIQUES *Sulla classificazione delle superficie algebriche e particolarmente sulle superficie di genere lineare* $p^{(1)} = 1$. Rend. Acc. Lincei, s. 5^a, **23** (1914).

[E5] F. ENRIQUES *Le superficie di genere uno.* Rend. Acc. di Bologna, **13** (1909).

[E6] F. ENRIQUES *Ricerche di geometria sulle superficie algebriche.* Memorie Acc. Torino, s. 2^a, **44** (1893).

[E7] F. ENRIQUES *Sopra le superficie algebriche di bigenere uno.* Memorie della Società dei XL, s. 3^a, **14** (1906).

[E8] F. ENRIQUES *Le superficie algebriche.* Zanichelli (1949).

[EGA] A. GROTHENDIECK *Eléments de géométrie algébrique.* Publ. Math. IHES. **4, 8, 11, 17, 20, 24, 28.**

[Ek] T. EKEDAHL *Canonical models of surfaces of general type in positive characteristic.* Publ. Math. IHES **67** (1988).

[FAC] J.-P. SERRE *Faisceaux algébriques cohérents.* Ann. of Math. **61** (1955).

[F-M] R. FRIEDMAN and J. MORGAN *Algebraic surfaces and Seiberg-Witten invariants.* To appear.

[F-Q] R. FRIEDMAN and Z. QIN *On complex surfaces diffeomorphic to rational surfaces.* Invent. Math. **120** (1995).

[G] P. GRIFFITHS *Variations on a theorem of Abel.* Invent. Math. **35** (1976).

[G-H] P. GRIFFITHS and G. HARRIS *Principles of Algebraic Geometry.* John Wiley (1978).

[GAGA] J.-P. SERRE *Géométrie algébrique et géométrie analytique.* Ann. Inst. Fourier **6** (1955–56).

[Gr] A. GROTHENDIECK *Familles d'espaces complexes et fondements de la géométrie analytique.* Exposés 7-8-16-17 in Séminaire Cartan 13 (1960–61).

[H] F. HIRZEBRUCH *Topological methods in algebraic topology.* Springer-Verlag (1956).

[I-M] V.A. ISKOVSKIH and Y. MANIN *Three-dimensional quartics and counterexamples to the Lüroth problem.* Mat. Sbornik **86** (1971).

[K] K. KODAIRA *On compact complex analytic surfaces II.* Ann. of Math. **77** (1963).

[M1] D. MUMFORD *Lectures on curves on an algebraic surface.* Princeton University Press (1966).

[M2] D. MUMFORD *Geometric invariant theory.* Springer-Verlag (1965).

[M3] D. MUMFORD *Abelian varieties.* Oxford University Press (1970).

[N1] M. NOETHER *Zur Theorie des eindeutigen Entsprechends algebraischer Gebilde.* Math. Annalen **2** (1870) and **8** (1875).

[N2] M. NOETHER *Extension du théorème de Riemann-Roch aux surfaces algébriques.* C. R. Acad. Sci. Paris **103** (1886).

[N3] M. NOETHER *Über Flächen, welche Schaaren rationaler Kurven besitzen.* Math. Ann. **3** (1870).

[P] C. PETERS *Introduction to the theory of compact complex surfaces.* Canad. Math. Soc. Conf. Proc. **12** (1992).

[R] M. RAYNAUD *Familles de fibrés vectoriels sur une surface de Riemann.* Sém. Bourbaki Exp. 316 (1966).

[S1] J.-P. SERRE *Un théorème de dualité.* Comment. Mat. Helv. **29** (1955).

[S2] J.-P. SERRE *Critère de rationalité pour les surfaces algébriques.* Sém. Bourbaki Exp. 146 (1956–57).

[Se1] F. SEVERI *Sulla classificazione delle rigate algebriche.* Rend. di Mat., s. 5, **2** (1941).

[Se2] F. SEVERI: *Intorno ai punti doppi impropri di una superficie generale dello spazio a 4 dimensioni e a suoi punti tripli apparenti.* Rendiconti di Palermo **15** (1901).

[Se3] F. SEVERI: *Le superficie algebriche con curva canonica d'ordine zero.* Atti del Ist. Veneto **68** (1909).

[Sg1] C. SEGRE *Recherches générales sur les courbes et les surfaces réglées algébriques.* Math. Ann. **30** (1887) and **34** (1889).

[Sg2] C. SEGRE *Etude des différentes surfaces du 4e ordre à conique double ...* Math. Annalen **24** (1884).

[Sh1] I.R. SHAFAREVICH *Basic algebraic geometry.* Springer-Verlag (1974).

[Sh2] I.R. SHAFAREVICH et al. *Algebraic Surfaces.* Proc. of the Steklov Institute **75** (1965).

[Sh-P] I.R. SHAFAREVICH and I. PJATETSKII-SHAPIRO *A Torelli theorem for algebraic surfaces of type K3.* Math. USSR Izvestia **5** (1971).

[V] G. VERONESE *La superficie omaloide normale a 2 dimensioni e del 4° ordine dello spazio a 5 dimensioni...* Memorie della R. Acc. dei Lincei, s. III, **19** (1884).

[Va] G. VACCARO *Le superficie razionali prive di curve eccezionali di 1^a specie.* Rend. Lincei, s. 8^a, **4** (1948).

[W] A. WEIL *Variétés Kählériennes.* Hermann (1958).

[Wi] E. WITTEN *Monopoles and four-manifolds.* Math. Research Letters **1** (1994).

[X1] M. DEMAZURE, H. PINKHAM and B. TEISSIER *Séminaire sur les singularités des surfaces (Palaiseau).* Lecture Notes in Math. **777**, Springer-Verlag (1980).

[X2] *Géométrie des surfaces K3: modules et périodes (Séminaire Palaiseau).* Astérisque **126** (1985).

[Z1] O. ZARISKI *Reduction of singularities of algebraic 3-dimensional varieties.* Ann. Math. **45** (1944).

[Z2] O. ZARISKI *The problem of minimal models in the theory of algebraic surfaces.* Amer. J. Math. **80** (1958).

[Z3] O. ZARISKI *On Castelnuovo's criterion of rationality* $p_a = P_2 = 0$ *of an algebraic surfaces.* Illin. J. Math. **2** (1958).

Index

Printed in the United States
By Bookmasters